FLEXIBLE COMPUTING IN ECOLOGY

FLEXIBLE COMPUTING IN STATISTICAL ECOLOGY

László Orlóci

Visiting Professor, Laboratory of Plant Quantitative Ecology, Department of Ecology, Universidade Federal do Rio Grande do Sul, Porto Alegre, Brazil

London, Canada 2011

Scada
Publishing

This Book is an external appendix of L. Orlóci's Statistical Ecology*. A second appendix contains the executable (.exe) computer programs. To receive a copy of the programs please contact: SCADA.LONDON@GMAIL.COM .

Reference:
Orlóci. L. 2011. Flexible Computing in Statistical Ecology.. Scada Publishing. Online Edition, CreateSpace (www.createspace.com/3574792).

Look for:
*Orlóci, L. 2010. Statistical Ecology. The quantitative exploration of nature to reveal the unexpected. Scada Publishing. Online Edition, Create-Space (www.createspace.com/3476529)

.

ISBN 978-1460972953

Contents

8

Preface

The reasoned approach to problem solving begins with isolation of the individual tasks that bear on the solution. How well can this be done? It depends on how much is already known about the problem and the solution. In the environmental studies, and especially on the macro level which global change studies are targeting, the information available initially for the researcher is usually very low. It is a consequence of this that problem solving in that field progresses through levels of approximation.

One of the pre-requisites for success is the versatility and task specificity of the analytical tools. I am presenting *my* tools: computer programs which I wrote as they were needed at different times from the 1960's to the present. These served my purposes well in that they target specific problems and when needed allow modifications to broaden their utility; hence the name 'flexible computing'. I never had time to make my computer programs usable without hands on instructions or without examples showing how to get the different tasks underway under a sometimes rather lengthy, interrogative start up dialogue. While the dialogue's contents make easy the mapping out the specific flow ofS the analysis, selection of options and answers to questions are involved which assume familiarity not only with the technical language but also with the fine details of the techniques about which decisions are being made. I found to teach these not too difficult in regular and short courses. But I have yet to find the novice to whom I would recommend to begin their data analytical carrier with the program. I therefore suggest as a first step a careful study of my presentation of the subject in "Statistical Ecology" written in 2010. The book is distributed by Amazon.

A successful study of "Statistical Ecology" requires three things: reading with complete attention and if necessary memorizing entire phrases and definitions; learning to express once thoughts in the technical language of the subject, avoiding past, cursory exposure to the subject to get in the way; learn to study with a writing block and pencil in hand.

What I try to do in the programs is to make them specific yet chainable. "Specific" implies that although the tools may be packaged, they are addressable individually and in the combinations which one's research interests dictate. Each program supports the computation of well defined task. The utility in this is most clear to the instructor of introductory courses and in front line research where full control over the flow of calculations is needed. Computer program should allow users to escape the suffocating straightjacket of the automated do-it-all canned routines. The research problem should never be redefined to fit the short coming of a canned statistical program.

The first set of the APICE (EPIC) programs began their evolution in the early '60s in ALGOL on the ELLIOTT 803 computer (paper tape input and output) at the University College of North Wales in Professor Peter Greig-Smith's laboratory. The programs continued to evolve through many revisions (1975, 1978, 1991, 1995). The last major revisions were implemented in 2009 synchronized with the exercises in Or-lóci's "Statistical Ecology" (2010). [1]

All programs are conversational. This implies an occasion-ally lengthy start up dialogue in the form of requests for in-formation and questions regarding the analytical path to be taken by the program and responses by the user. This is a feature which encourages pre-analysis study of the meth-ods and familiarity with the data. The start up dialogue does in fact permit users to select the tasks to be performed and to control the volume of output from any run. This facilitates access to details that the user may wish to see to checks on his or her long-hand calculations -- a must for understand-ing statistics. As a drawback, the dialogue will present diffi-culties for new users. This can be helped by preparatory studies and by putting the programs through their paces in trial analyses.

The programs are presented as free-standing applications. They are fully synchronised with the exercises in Orlóci's "Statistical Ecology" (2010). It will be good practice to have this manual and also the book in hands when trying out the programs. The code in the very early versions brings up a DOS widow and the conversation takes place within it. The usual DOS restrictions on file names apply. So a name EX41221.txt is correct but EX41221Cov.txt is not. This is because 8 characters are allowed in the name before the extension name. Should when a run fails, a file name or file content offense may be the offending problem. The more recent versions of the programs are presented in compiled and bound (linked) True Basis. All programs were compiled and linked on a 32 bit Windows XP system. Tests indicate that they will run on Windows Vista and Windows 7. It is

---

[1]A phase of the process in the 1980's involved the help of N.C. Kenkel who contributed BASIC code for the programs REGRESSION, ANOVA, FANOVA and MDSCAL.

strongly advised to work in isolated WORK folders placed on or near the top of the directory (c:\WORK).

The required minimum internal memory depends on the code and on the size of the arrays. The arrays are dynamically dimensioned, but low memory as the internal files are generated will certainly stop the processing.

The manual has three parts. The first describe data entry, the second the mechanics of running the programs, and the third presents the programs themselves within groups. The third part includes program details, and specifications regarding input and output. The runs produce printable disk files. The newer code produces graphics files as well.

L. Orlóci

Porto Alegre, Winter 2009

1 DATA ENTRY

1.1 Locating files

All programs require some data entry through the keyboard during the run. Most programs require also mass data entry from disk. File name and location are among the information items requested by the programs in the start-up dialogue. A typical file designation may take the form of

C:\WORKS\ex41221.TXT

This name locates a data file named ex41221 in the folder named WORKS on the hard drive (HD, volume) C. If both the data and program are in the same folder the same file, the program will find the file by the name:

ex41221.txt

Upper or lower case types are accepted. The file named happens to be Example 4.1.2.2.1 in the companion book "Statistical Analysis".

It is good practice to collect program and data files in a common WORKS folder on hard disk in the uppermost directory segment with ample free space left on the hard drive itself. Executable programs and data files are included on the attached media.

1.2 File format

The data format must conform to the format expected by the program. The data files may contain numerals or symbols of the Alphabet. The data are expected to be in one of the standard formats. For example, the 3 by 5 matrix

```
2  3  1   0  5
8  4 40   6 -9
0  0  4  10  2
```

will have a text file image:

```
2
3
1
0
5
8
4
40
6
-9
0
0
4
10
2
```

The above arrangement is called "entry by matrix rows". The "Entry by matrix column" looks like this:

```
2
8
0
3
4
0
1
40
```

4
0
6
10
5
-9
2

From the programming point of view, the sole purpose of terms like "row" or "column" is to keep track of "row" or "column" specific operations and output by position without engaging an elaborate labelling system. Input and output files are text (.TXT or .txt) files.

Three types of primary data arrangements are distinguished:

Arrangement 1 or full matrix arrangement. The forgoing examples are typical for this mode of data presentation.

Arrangement 2 or half matrix arrangement. Data from the upper half of a symmetric matrix are entered by row, each row beginning with the value in the principle diagonal cell. For example, given the 3 x 3 matrix:

0	0.2	1.9
0.2	1	0.5
1.9	0.5	3

the numbers entered in arrangement 2 are:

0

0.2

1.9

1

0.5

3

Arrangement 3 or single vector arrangement. This is the same as Arrangement 1, except the grouping of unrelated vectors is involved in a single file. For example, given the 4-valued vector,

  3   1   6   19

and a 3-valued vector

  82   5   17

the data arrangement 3 type entry is:

3
1
6
19
82
5
17

1.3 Program created files

A program may output the data file for input in another program or for viewing. In any case the output format is automatic.

1.4 User created data files

After entry on the editor's page, select SAVE AS from the file menu, key-in a volume name and/or a file name and click SAVE. Do not use leading or trailing blanks in file names! Make the names short and descriptive. Observe the 8 character limitation in some programs. Always create a backup file on a different disk under a different file name (e.g., ex41221.BAK). Always check that the save commend actually saves the file to the folder intended.

1.5 Important note

The examples are set up for long-hand verification in most cases. This makes it necessary to rely upon small data sets notwithstanding the logical requirements of the problem that may suggest validity of the statistics computed only when the sample size (data set) is exceptionally large. Expect a very reach printed output content from the programs. Go easy on hard copies.

2 THE .exe PROGRAMS

2.1 Initial preparations and start-up dialogue

It is good practice before starting data analysis to study the appropriate section in "Statistical Ecology". Special attention should be paid to the example to be recomputed. The terminology should be kept and used as given. It is also good practice to place program and data file(s) into a WORKS folder. Start program by clicking on the program's disk icon. The output from some programs is a matter of a few numbers which can be copied longhand. If the output is more voluminous, it is directed automatically to a disk file, usually under one of the file names:
PRINTDA.TXT , Printda.TRU

The start-up dialogue is important and the user responses must be precise and timely. Users must be familiar not just with the meaning of the technical terms (see Glossary in "Statistical Ecology") but also with the entire method itself. The responses will determine the outcome. No default decisions are programmed!

2.2 The PRINTDA file

Most results and vital information about the run are stored in the run's specific output file PRINTDA.TXT. This file is written by the program into the folder where the input data file is located under an extended name. The file name extension is usually keyed in by the user when so requested in the start-up dialogue. It is prudent to use a short descriptive extension name to further specify the PRINTA core file name. For example: PRINTDA1.TXT .

The PRINTDA file can be opened in a word processing program, format-adjusted if needed, and printed.

2.3 Capturing screen objects

Not all information on the screen are captured and written on file by the program. The exact content of the start up dialogue is this kind. Any information on the screen not captured will be lost when the screen is cleared. To capture screen contents follow your system's conventions. In Windows pressing the [Prt Scr] key or in lieu of this, pressing a specified key combination will copy the screen to the Clipboard. The copy can be pasted in some paint text editor program and saved on permanent file.

3 EXAMPLES

The programs are grouped in sections according to generic type. The descriptions are structured:

Part 1: brief characterisation.
Part 2: data file identification.
Part 3: PRINTDA file listing, showing input/output file names, other information, and other results of the run.
Part 4: graphics screen, if any, etc..

3.1 Capturing elements of the data structure

3.1.1 Program ANGLES

P1: Inner angles are computed for lines connecting the tips of vectors ordered in n-dimensional analytical space. The vectors may represent relevés. The vector tips are arranged on a transect as A, B, C, … . For example, if

    A B C

represent relevé vectors with tips

    a b c

the inner angle computed is for lines a,b and b.c at b. Data entry on file is by variable (Arrangement 1). Angles are stored on user-specified file.

P2:   EX41221.txt Starfish counts in tidal pools.

Species				Tidal pools			
Heliaster kubiniji	26	29	29	29	30	35	39
Pisaster ochraceaus	28	31	31	33	27	38	36
Pisaster brevispinus	18	14	13	13	19	15	15

Data entered on file as:
26
29
29
29
30
35
39
28
31
31
33
27
38
36
18
14
13
13
19
15
15

P3:   PROGRAM ANGLES PRINTDA FILE

COORDINATES FILE: ex41221.txt
ANGLES FILE: angles.txt
NUMBER OF AXES: 3
NUMBER OF UNITS: 7
OUTPUT IN FOLLOWING ORDER:
POSITION ON TRANSECT, DISTANCE AB, DISTANCE AC, DISTANCE BC, COS, ANG

2:	34.00	43.00	1.00-	.69	133.31
3:	1.00	5.00	4.00	.00	90.00
4:	4.00	53.00	73.00	.70	45.39
5:	73.00	65.00	162.00	.78	38.59
6:	162.00	178.00	20.00	.04	87.99

Time: 14:23:07
Today is 09-09-08

3.1.2 Program CALHOUN

P1: Calhoun's set theoretical distance is computed for relevé pairs. Data input is by variable (Arrangement 1). Half of the distance matrix is written onto an Arrangement 2 file.

P2: EX41221.txt (see table under section 3.1.1)

P3:   PROGRAM CALHOUN
=========================
COORDINATES FILE: ex41221.txt
DISTANCES FILE: distances.txt
NUMBER OF VARIABLES: 3
NUMBER OF POINTS: 7

CALHOUN DISTANCES MATRIX (LOWER HALF AND DIAGONAL CELLS):
```
  0
 18   0
 21   3   0
 24   3   3   0
 18  24  27  30   0
 30  21  24  21  30   0
 30  21  24  21  30   0   0
```

Time: 15:34:57
Today is 09-09-08

3.1.3 Program CORRELATION

P1: Product moments and the duals of productmoments are computed for pairs of variables or relevés. Typically the co-variance involves variables while the dual of the covariance involves relevés, both using the same variable-centred data set. The type of product moment is option controlled. Data entry is by variable from an Arrangement 1 file. Products are stored in an Arrangement 2 file.

P2:   EX 41221.txt (SECTION 3.1.1)

P3:   PROGRAM CORRELATION

ROGRAM CORRELATION
COORDINATES FILE:ex41221.txt, VOLUME
PRODUCTS MATRIX. COVARIANCE (Option 1)
DUAL MATRIX CALCULATED !   !!

```
  8.0612
  1.7517    1.1088
```

```
  1.2993    1.3231    1.7041
  0.0340     .9898    1.3707    1.7041
  5.8469     .3707-    .2483-   1.9150    6.6327
- 7.4626-   2.2721-   2.2245-    .2245-   5.8435  8.6803
- 9.4626-   3.2721-   3.2245-   1.8912-   4.8435  9.3469  13.3469
```

PRODUCTS FILE: qcov.txt
Time: 12:52:52
Today is 09-09-09

3.1.4 Program GENDIST

P1: The Mahalanobis generalised distance is computed for given groups and one or more external assignable objects (relevé vectors) based on a common set of variables. Two data files are required: one for the groups (Arrangement 3) and one for the objects to be assigned (Arrangement 1). Data entry is by relevé. The distances are stored in an Arrangement 2 file under user specified name.

P2: EX1821gr.txt Species densities in two vegetation types:

Species	Type 1 (n1=6)					
Pseudotsuga menziesii	10	21	0	12	30	0
Acer macrophyllum	16	17	3	18	18	6
Thuja plicata	20	21	7	21	34	2

	Type 2 (n2=6)					
Pseudotsuga menziesii	10	30	10	12	28	11
Acer macrophyllum	10	41	4	11	43	8
Thuja plicata	6	55	9	8	20	12

Data entered on file as:
10
21
0
12
30
0
16
17
3
18
18
6
20

21
7
21
34
2
10
30
10
12
28
11
10
41
4
11
43
8
6
55
9
8
20
12

        EX1821X.txt
Relevé vector to be assigned:
Pseudotsuga menziesii   10
Acer macrophyllum   30
Thuja plicata   10

Data entered on file as:
10
30
10

P3: PROGRAM: GENDIST
GROUPS FILE: ex1821gr.txt
NUMBER OF GROUPS: 2
NUMBER OF VARIABLES: 3
NUMBER OF ASSIGNABLE OBJECTS: 1
GROUP SIZES: 6 6
EXTERNAL ASSIGNABLE OBJECTS FILE:: ex1821x.txt
EXTERNAL OBJECT(S):
OBJECT 1 : 10.000 30.000 10.000
    === GROUP 1 ===
COVARIANCE MATRIX
   19.367   4.100   9.400
    4.100 47.500- 9.800
    9.400- 9.800   8.800

MEAN VECTOR
   16.167 23.500   3.000

INVERSE OF COVARIANCE MATRIX

```
  .283695-  .112963-  .428838
- .112963   .072313   .201196
- .428838   .201196   .795773
```

ASSIGNMENT OF OBJECT 1
GENERALIZED DISTANCE =   10.8270
F-VALUE = 140.669
NUMERATOR DEGREES OF FREEDOM = 3
DENOMINATOR DEGREES OF FREEDOM = 3
     === GROUP 2 ===
COVARIANCE MATRIX
```
  4.700- 10.500-  .600
- 10.500 158.167- 20.400
- .600- 20.400  8.000
```

MEAN VECTOR
```
  9.500   36.167  9.000
```

INVERSE OF COVARIANCE MATRIX
```
  .303804   .034431   .110585
  .034431   .013323   .036556
  .110585   .036556   .226513
```
ASSIGNMENT OF OBJECT 1
GENERALIZED DISTANCE =   .5065
F-VALUE = .308
NUMERATOR DEGREES OF FREEDOM = 3
DENOMINATOR DEGREES OF FREEDOM = 3
Time: 14:01:16
Today is 09-09-09

Assign X to group for which its generalised distance is smallest if the observed F is small (its probability of occurring by chance is large). Use program FPROBS to compute the probability.

3.1.5 Program METRICS

P1: Distances are computed between pairs of individuals. The raw data are read from an Arrangement 1 file. Input is by variables. Distances are stored in an Arrangement 2 file.

P2: EX1511.txt Tree densities in plots laid on an elevation gradient at 200 m intervals.

Elevation	LO																			HI
Plot	1	2	3	4	5	6	7	8	9	10	11	12	13	14	15	16	17	18	19	20
A	40	45	50	47	41	35	25	20	15	13	10	7	5	2	2	2	0	0	0	0
B	10	16	22	37	40	45	42	40	36	30	25	20	16	13	7	3	2	1	0	0
C	0	2	5	11	13	17	25	31	41	51	58	60	51	40	35	28	25	20	14	10
D	0	0	1	1	1	2	5	8	12	15	19	26	35	52	60	66	60	49	36	30

Data entered on file as:
40
45
50
47
41
35
25
20
15
13
10
7
5
2
.
.
.
8
12
15
19
26
35
52
60
66
60
49
36
30

P3:
PROGRAM: METRICS
=========================
DATA FILE:ex1511.txt
NUMBER OF VARIABLES = 4
NUMBER OF OBSERVATIONS PER VARIABLE = 20
DISTANCE OPTION (1): ABSOLUTE VALUE FUNCTION

DISTANCES FILE:
0 13 28 46 45 59 77 89 104 113 122 129 127 133 136 139 133 118 100 90
0 15 33 40 56 74 86 101 110 119 126 124 136 145 148 142 127 109 99
0 24 35 51 69 81 96 105 114 125 131 143 152 155 149 134 116 106
0 11 27 45 57 74 95 114 131 137 149 158 161 155 140 122 114
0 16 34 46 69 90 109 126 132 144 153 156 150 135 117 113
0 24 40 63 84 103 120 126 138 147 150 144 129 117 115
0 16 39 60 79 96 102 114 123 126 120 115 109 107
0 23 44 63 80 86 98 107 116 116 111 105 103
0 21 40 57 63 77 96 113 113 108 102 100
0 19 36 42 76 95 112 112 107 101 99
0 17 37 71 90 107 107 102 96 94

```
0 24 58 77 94 94 89 83 81
0 34 53 70 70 65 59 67
0 19 36 36 37 57 67
0 17 17 34 54 64
0 12 29 49 59
0 17 37 47
0 20 30
0 10
0
```

DISTANCES FILE: AVF.txt
Time: 15:28:56
Today is 09-09-09

3.1.6 Program PINDEX

P1: A version of Goodall's probabilistic similarity index is computed for pairs of relevés from data stored on an Arrangement 1 file. Data entry is by variable. The similarity values are stored in an Arrangement 2 file.

P2: EX41221.txt (Section 3.1.1)

P3: PROGRAM PINDEX
=======================
COORDINATES FILE:EX41221.txt, VOLUME C:\QB45\WORKS
NUMBER OF VARIABLES: 3
NUMBER OF OBSERVATIONS PER VARIABLE: 7

DIFFERENCE VECTORS

PAIR LABELS AND DIFFERENCES ARE PRINTED

```
VARIABLE 1:
 1 2 3.0 1 3 3.0 1 4 3.0 1 5 4.0 1 6 9.0 1 7 13.0
 2 3 0.0 2 4 0.0
 2 5 1.0 2 6 6.0 2 7 10.0 3 4 0.0
 3 5 1.0 3 6 6.0 3 7 10.0 4 5 1.0
 4 6 6.0 4 7 10.0
 5 6 5.0 5 7 9.0 6 7 4.0
VARIABLE 2:
 1 2 3.0 1 3 3.0 1 4 5.0 1 5 1.0 1 6 10.0 1 7 8.0
 2 3 0.0 2 4 2.0
 2 5 4.0 2 6 7.0 2 7 5.0 3 4 2.0
 3 5 4.0 3 6 7.0 3 7 5.0 4 5 6.0
 4 6 5.0 4 7 3.0
 5 6 11.0 5 7 9.0 6 7 2.0
VARIABLE 3:
 1 2 4.0 1 3 5.0 1 4 5.0 1 5 1.0 1 6 3.0 1 7 3.0
 2 3 1.0 2 4 1.0
```

```
2 5 5.0 2 6 1.0 2 7 1.0 3 4 0.0
3 5 6.0 3 6 2.0 3 7 2.0 4 5 6.0
4 6 2.0 4 7 2.0
5 6 4.0 5 7 4.0 6 7 0.0
```

PAIR	PROBABILITY VECTOR
COMPARED	FOR VARIABLES

```
1 2 0.4286 0.3810 0.7619
1 3 0.4286 0.3810 0.9048
1 4 0.4286 0.6667 0.9048
1 5 0.5238 0.0952 0.2381
1 6 0.8095 0.9524 0.6190
1 7 1.0000 0.8571 0.6190
2 3 0.1429 0.0476 0.3333
2 4 0.1429 0.2381 0.3333
2 5 0.2857 0.4762 0.8095
2 6 0.7143 0.8095 0.2381
2 7 0.9524 0.6190 0.2381
3 4 0.1429 0.2381 0.0952
3 5 0.2857 0.4762 1.0000
3 6 0.7143 0.8095 0.5238
3 7 0.9524 0.6190 0.5238
4 5 0.2857 0.7143 1.0000
4 6 0.7143 0.5238 0.5238
4 7 0.9524 0.2857 0.5238
5 6 0.5714 1.0000 0.7619
5 7 0.7619 0.9048 0.7619
6 7 0.4762 0.1429 0.0476
```

PAIR	CHI	ALPHA	PROBABILITY
COMPARED	SQUARE	PROB.	INDEX (S)

```
1 2 4.1686 0.6539 0.3461
1 3 3.8249 0.7004 0.2996
1 4 2.7057 0.8448 0.1552
1 5 8.8662 0.1812 0.8188
1 6 1.4793 0.9609 0.0391
1 7 1.2674 0.9734 0.0266
2 3 12.1781 0.0581 0.9419
2 4 8.9592 0.1759 0.8241
2 5 4.4120 0.6211 0.3789
2 6 3.9657 0.6813 0.3187
2 7 3.9269 0.6866 0.3134
3 4 11.4647 0.0750 0.9250
3 5 3.9894 0.6781 0.3219
3 6 2.3888 0.8807 0.1193
3 7 2.3500 0.8849 0.1151
4 5 3.1785 0.7861 0.2139
4 6 3.2595 0.7756 0.2244
4 7 3.8964 0.6907 0.3093
5 6 1.6631 0.9479 0.0521
```

5 7 1.2879 0.9723 0.0277
6 7 11.4647 0.0750 0.9250
SIMILARITY VALUES IN FILE:C:\QB45\WORKS\SIM

3.2 Measuring commonness

3.2.1 Program CHIPROBS

P1: The algorithm uses polynomial approximation to compute a Chi-squared value when probability alpha is given or a probability when a chi-squared value is given. The known (chi-squared or alpha, and the degrees of freedom) are specified by user in keyboard input. Results should be copied from the screen.

P2: No disk file entry.
P3: No PRINTDA file created.

3.2.2 Program DISTRIBUTIONS

P1: The Poisson distributions is fitted to a sample distribution. Data entry is from an Arrangement 1 file. The fitted frequencies and goodness-of-fit measure are stored in a PPRINTDA file.

P2: EX42111.txt Clutch size frequencies in Turdus migratorius nests.

X	0	1	2	3	4	>4
f(X)	6	15	19	14	7	1

Data entered on file as:
6
15
19
14
7
1

P3: PROGRAM: DISTRIBUTIONS
==============================
DISTRIBUTION FILE: EX42111.TXT, VOLUME
NUMBER OF CATEGORIES: 6
OPTION: POISSON DISTRIBUTION.
MEAN = 2.064516
SECOND MOMENT = 1.447451

```
--------------------------------------------------
     OBSERVED DATA     FITTED VALUES
  X   P(X)  F(X)     P'(X)   F'(X)
--------------------------------------------------
  0  0.0968  6      0.1269   7.8665
  1  0.2419  15     0.2619  16.2406
  2  0.3065  19     0.2704  16.7645
  3  0.2258  14     0.1861  11.5369
  4  0.1129  7      0.0960   5.9545
  5  0.0161  1      0.0587   3.6370
--------------------------------------------------
```

NOTE: VALUES AT X= 5 INCLUDE THOSE GREATER THAN 5
TEST FOR GOODNESS OF FIT:

2I (INFORMATION DIVERGENCE) = 4.223118
CHI-SQUARE = 3.457183
DEGREES OF FREEDOM = 4

3.2.3 Program ENTRDIST

P1: An empirical distribution is generated for interaction (mutual) information. The technique uses unrestricted randomisation with Monte Carlo simulation. The constants are input from keyboard. The information values are stored in ascending order in the PRINTDA file. Probabilities can be inferred from position in the ordered output.

P2: Keyboard input of constants describing the contingency table characteristics in EX6321.txt

Tree species   1112221123322331 1122233311221121
Yeast species   3234443341134213 3343411233433342

P3: PROGRAM ENTRDIST
=============================
NUMBER OF STATES OF VARIABLE 1: 3
NUMBER OF STATES OF VARIABLE 2: 4
SAMPLE SIZE = 32
NUMBER OF ITERATIONS: 100

SORTED INFORMATION VALUES:

```
  1  0.02966  2  0.03060  3  0.03564  4  0.03573  5  0.04043
  6  0.04291  7  0.04313  8  0.04397  9  0.04444 10  0.04497
 11  0.04592 12  0.04820 13  0.04967 14  0.04983 15  0.05161
 16  0.05302 17  0.05302 18  0.05315 19  0.05641 20  0.05831
```

21	0.06060	22	0.06123	23	0.06220	24	0.06284	25	0.06312
26	0.06354	27	0.06928	28	0.06981	29	0.07019	30	0.07156
31	0.07176	32	0.07339	33	0.07347	34	0.07386	35	0.07577
36	0.07846	37	0.07878	38	0.07911	39	0.08040	40	0.08234
41	0.08256	42	0.08379	43	0.08397	44	0.08401	45	0.08435
46	0.08619	47	0.08811	48	0.08820	49	0.09026	50	0.09074
51	0.09153	52	0.09557	53	0.09590	54	0.09960	55	0.09986
56	0.10145	57	0.10484	58	0.10489	59	0.10979	60	0.11072
61	0.11416	62	0.11416	63	0.11958	64	0.12043	65	0.12648
66	0.12711	67	0.12816	68	0.12880	69	0.13674	70	0.13795
71	0.13858	72	0.14051	73	0.14446	74	0.14506	75	0.14873
76	0.15119	77	0.15721	78	0.16090	79	0.16366	80	0.16394
81	0.16436	82	0.17055	83	0.17105	84	0.18649	85	0.18966
86	0.19319	87	0.19489	88	0.19796	89	0.19949	90	0.20278
91	0.20389	92	0.21488	93	0.22083	94	0.23843	95	0.24739
96	0.25967	97	0.28182	98	0.28466	99	0.30171	100	0.30206

AVERAGE INFORMATION (BIAS) = 0.11402
VARIANCE = 1.345586E-04

Interpretation:
1. The bias to be expected in the estimation of mutual information in this case is 0.114. Thus the unbiased estimate in Example 6.3.2 is 0.752-0.114=0.638 .
2. The probability that under the rule of chance an I value will exceed 0.752 is very small <<1/100. This proportion is valid for any 32 totalled 12-valued contingency table.

3.2.4 Program FPROBS

P1: Polynomial approximation is used to compute an F-value when a probability is given, or a probability when an F-value is given at specified numerator and denominator degrees of freedom. Entry of the three parameters is from the keyboard. Results should be copied from the screen.

P2: All input through the keyboard.
P3:   No PRINTDA file created.

3.2.5 Program NPROBS

P1: Polynomial approximation is used to compute a Z-value when a probability is given, or a probability when a Z value is given. Z is the standard normal variate. The given (Z or probability) is entered through the keyboard. Results should be copied from the screen.

P2: All input through the keyboard.
P3:   No PRINTDA file created.

3.2.6 Program TPROBS

P1: The technique is polynomial approximation. The output is a t-value when a probability is given or a probability when a t-value is given. Entry of t-value or probability, and the degrees of freedom is from the keyboard. The results appear on the screen.

P2: All input through the keyboard.
P3: No PRINTDA file created.

3.3 Describing distributions

3.3.1 Program ENTROPY

P1: Entropy and information quantities are computed. Interactions involve pairs of variables. Any number of variables are permitted. The data file is Arrangement 1 entered by variable. The data elements are taken as if they were diagnostic labels, unordered, coded as numerals. Sample size must be large.

P2: ex41331.txt Classification of plants in a stand between 3 life-forms and 4 strata.
Life-form X1   111111111112222 222222233333444
Stratum X2   111112223331222 222222333333333

Data entered on file as:
1
1
1
1
1
1
1
1
1
1
1
2
2
2
.
.
.

2
3
3
3
3
3
3
3
3
3

P3: PROGRAM: ENTROPY
========================
DATA FILE: ex41331.txt, VOLUME c:\qb45\entropy
NUMBER OF VARIABLES: 2
SAMPLE SIZE (NUMBER OF OBJECTS OBSERVED): 30
STATES OF VARIABLE 1
----
 1 2 3 4
FREQUENCIES OF THESE STATES
----
 11 11 5 3
ENTROPY FUNCTIONS
----
1. SHANNON ENTROPY = 1.26464
2. LOG SIMPSON INDEX = 1.181994
(MAXIMUM OF THESE = 1.386294)
STATES OF VARIABLE 2
----
 1 2 3
FREQUENCIES OF THESE STATES
----
 6 12 12
ENTROPY FUNCTIONS
----
1. SHANNON ENTROPY = 1.05492
2. LOG SIMPSON INDEX = 1.021651
(MAXIMUM OF THESE = 1.098612)

JOINT FREQUENCIES, VARIABLES 1 AND 2
----
  5 3 3
  1 9 1
  0 0 5
  0 0 3

JOINT ENTROPY = 1.875967
MUTUAL ENTROPY (INFORMATION) = .4435931 (MAXIMUM = 1.098612)
RAJSKI'S METRIC = .763539
COHERENCE COEFFICIENT = .6457618

3.3.2 Program ENTROGRA

P1: Graphs are drawn for Rényi's entropy of order alpha. Lower limit of order must be given as zero or a number greater than zero, but not exactly one. Any number of distributions are permitted and the distributions may have different numbers of elements and different totals. Data entry is by distribution (Arrangement 3 file). The graphs drawn on screen are captured automatically in PICT files.

P2: ex41321.txt (see Section 3.3.1)

P3: PROGRAM EntroGraphs

Entropy of order alpha is computed and entropy graphs drawn for F and the limiting Fm and Fl.

Lower limit alpha= 0
Upper limit alpha= 50
Input data file name: ex41321.txt
DISTRIBUTION 1
PICT file: keptxt1.bmp
Maximum entropy: 1.0986

---

alpha	H alpha	minimum	evenness
.0000	1.0986	1.0986	1.0000
1.0000	.8135	.1568	.7405
2.0000	.6602	.0657	.6009
3.0000	.5785	.0504	.5266
4.0000	.5322	.0450	.4844
5.0000	.5038	.0423	.4586

Etc. up to alpha=50 (see graph)
PICT file: keptxt1.bmp
Maximum entropy: 1.0986

```
Data file: ex41321.txt
Distribution: 1
Alpha: 0  to  50
Maximum H= 1.0986123
Graphs: max H (top); H; min H
PRESS ANY KEY TO CONTINUE:
```

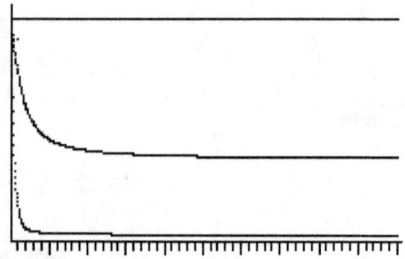

DISTRIBUTION 2
PICT file: keptxt2.bmp
Maximum entropy: 1.0986

alpha	H alpha	minimum	evenness
.0000	1.0986	1.0986	1.0000
1.0000	1.0986	.5481	1.0000
2.0000	1.0986	.3391	1.0000
3.0000	1.0986	.2704	1.0000
4.0000	1.0986	.2420	1.0000
5.0000	1.0986	.2273	1.0000

etc. up to alpha =50

```
Data file: ex41321.txt
Distribution: 2
Alpha: 0  to  50
Maximum H= 1.0986123
Graphs: max H (top); H; min H
PRESS ANY KEY TO CONTINUE:
```

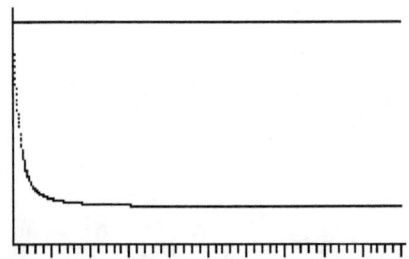

DISTRIBUTION 3
PICT file: keptxt3.bmp
Maximum entropy: 1.0986

alpha	H alpha	minimum	evenness
.0000	1.0986	1.0986	1.0000
1.0000	.5481	.5481	.4989
2.0000	.3391	.3391	.3087
3.0000	.2704	.2704	.2461
4.0000	.2420	.2420	.2203
5.0000	.2273	.2273	.2069

etc. up to alpha =50

38

```
Data file: ex41321.txt
Distribution: 3
Alpha: 0  to  50
Maximum H= 1.0986123
Graphs: max H (top); H; min H
PRESS ANY KEY TO CONTINUE:
```

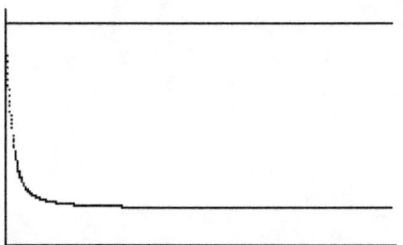

Graph information: entropy is plotted on vertical axis. The horizontal axis for alpha is calibrated from 0 to 50. Horizontal line on top is maximum entropy. Lower curve represents entropy in the least dispersed limiting distribution of the case.

3.3.3 Program ESTH

P1: Entropy is estimated from cumulative frequencies. The entropy quantity is Brillouin's and the method is Pielou's. Data file type is Arrangement 1 and data entry is by distribution (relevé). Sample size (the number of relevés) must be large for improved accuracy of the estimates.

P2: ex621t.txt Plant counts of three species in 6 quadrats.

Species		Quadrats				
1	100	93	43	87	97	97
2	1	26	42	65	100	86
3	27	50	11	17	21	19

The data entry on disk file is by column:
100
1
27
93
26
50
43
42

11
87
65
17
97
100
21
97
86
19

P3:
PROGRAM EST/H
==============================
NPUT DATA FILE: ex621t.txt, VOLUME
NUMBER OF DISTRIBUTIONS: 6
NUMBER OF ELEMENTS PER DISTRIBUTION: 3
CUMULATIVE COUNTS (VARIABLES):

100	193	236	323	420	517
1	27	69	134	234	320
27	77	88	105	126	145

CUMULATIVE COLUMN (INDIVIDUAL) TOTALS:

| 128 | 297 | 393 | 562 | 780 | 982 |

H-ASTERISC VALUES ARE:

| .5321 | .8297 | .9317 | .9623 | .9804 | .9786 |

ESTIMATES OF H (FROM INFLECTION POINT UP):

| 1.0335 | 1.0273 | .9713 |

CALCULATION OF H-HAT VALUE:

INFLECTION POINT: 3
MEAN H (UNBIASED ESTIMATE OF H) =   1.0107
MAXIMUM H =   1.0986
VARIANCE OF H-ESTIMATE =   .0004

3.3.4 Program MOMENTS

P1: Moments, product moments and shape constants (gamma values) are computed for variables from data stored in an Arrangement 1 file. Data entry is by variable. All results are saved in the PRINTDA file.

P2: ex41221.txt (see Section 3.1.1)
P3: PROGRAM MOMENTS
=========================
COORDINATES FILE: ex41221.txt, VOLUME c:\qb45\works
NUMBER OF VARIABLES: 3

NUMBER OF OBSERVATIONS PER VARIABLE: 7

VARIABLE MEANS

 31.0000 32.0000 15.2857
PRODUCT-MOMENTS MATRIX
(DIAGONALS: SECOND MOMENTS, OFF-DIAGONALS: PRODUCT MOMENTS)

 16.8571 11.8571 -1.2857
 11.8571 13.7143 -4.4286
 -1.2857 -4.4286 4.7755

COVARIANCE MATRIX
 (DIAGONALS: VARIANCES, OFF-DIAGONALS: COVARIANCES)

 19.6667 13.8333 -1.5000
 13.8333 16.0000 -5.1667
 -1.5000 -5.1667 5.5714

PRODUCT MOMENT CORRELATION MATRIX

 1.0000 0.7798 -0.1433
 0.7798 1.0000 -0.5472
 -0.1433 -0.5472 1.0000

THIRD MOMENTS

 60.8571 12.8571 6.4548

FOURTH MOMENTS
=--
 718.0000 348.0000 43.1345

3.3.5 Program NORMAL

P1: Normal densities are fitted to sample frequencies ($F(X)$) read from an Arrangement 3 file. The user must be prepared to key in the 1st class midpoint, the constant class interval, and the number of classes when requested by the program.

P2: ex42311.txt

Class i	Class limits m	Midpoint X_i	Frequency f(X)
1	45.5 - 50.5	48	5
2	50.5 - 55.5	53	11
3	55.5 - 60.5	58	23
4	60.5 - 65.5	63	31
5	65.5 - 70.5	68	10
6	70.5 - 75.5	73	1
			$f_. = 81$

Data entered as:
5
11
23
31
10
1

P3: PROGRAM NORMAL
=======================
FREQUENCIES FILE: EX42311.txt, VOLUME
NUMBER OF CLASSES: 6
MIDPOINT OF THE 1ST CLASS: 48
CONSTANT CLAS INTERVAL: 5

SAMPLE VALUES:

MEAN = 60.03704
SECOND MOMENT = 30.10974
THIRD MOMENT = -58.19749
FOURTH MOMENT = 2508.263
SCALE FACTOR (B) USED IN FITTING = 29.4

TEST OF SKEWNESS:

GAMMA 1 = -.3522441
SKEWNESS TO THE LEFT IS INDICATED.

TEST OF KURTOSIS:

GAMMA 2 = -.2333191
PLATYKURTIC CURVE IS INDICATED.

CLASS INTERVAL	OBSERVED FREQUENCY	FITTED DENSITIES
45.5 - 50.5	5	2.65
50.5 - 55.5	11	12.92
55.5 - 60.5	23	27.44
60.5 - 65.5	31	25.41
65.5 - 70.5	10	10.26
70.5 - 75.5	1	1.81

TEST FOR GOODNESS OF FIT:

2I (INFORMATION DIVERGENCE) = 5.318571
CHI-SQUARED = 4.679772
DEGREES OF FREEDOM = 5

3.4 Probabilistic comparisons

3.4.1 Program ANOVA

P1: Variants of the single factor univariate ANOVA are performed on data according to design:

> Option 1 - Complete randomized.
> Option 2 - Randomized blocks.
> Option 3 - Latin Square.

Option 1 allows unequal replicates per treatment. Options 2 and 3 assume the same number of blocks within each treatment and a single measurement in each of the blocks. Data entry is by treatment: all replicates of treatment A first, all replicates of treatment B second, etc. Option 3 requires two input files: the measurements file in which the numbers are entered by treatment (same as in Options 1 and 2), and a map (design) file whose elements identify the location of the measurements in the Latin Square. For example, given

> 36 67 40 43

the replicates of treatment A, the first set of numbers of the map file

> 2 4 3 1

identify 36 as the value in the 1st column 2nd row of the Latin square, 67 as the value in the 2nd column 4th row, etc. Given

> 40 36 47 65

the 2nd set in the measurements file (treatment B), and the 2nd set of numbers in the map file are given such as

> 1 3 2 4,

the Latin Square locations of the measurements are as follows: 40 in column 1 row 1, 36 in column 2 row 3, 47 in column 3 row 2, and 65 in column 4 row 4 .

A critical F probability point corresponding to given probability (alpha) and given numerator and denominator degrees of freedom (determined in the program) has to be keyed in when the program requests to complete the run.

P2a: ex113111.txt

Fertilizer 1	2	3	
Yield	34	40	44
	37	39	49
	40	37	43
	36		50
	37		

Data entered as:
34
37
40
36
37
40
39
37
44
49
43
50

P2b: ex11322.tx

Animals 1	2	3	
Groups	14	16	18
	17	19	22
	12	16	17
	15	16	18

Data entered as:
14
17
12
15
16
19
16
16
18
22
17
18

P2c   ex11332d.txt

	1	2	3	4
1	36	40	32	43
2	67	36	30	22
3	40	47	61	32
4	43	65	23	26

Data entered as:
36
67
40
43
40
36
47
65
32
30
61
23
43
22
32
26

ex11331m.txt

Treatment	Rows			
A	2	4	3	1
B	1	3	2	4
C	3	1	4	2
D	4	2	1	3

Data entered as:
2
4
3
1
1
3
2
4
3
1
4
2
4
2
1
3

P4a: PROGRAM ANOVA
=======================
OPTION 1: COMPLETE RANDOMIZED DESIGN
 MEASUREMENTS FILE: ex113111.txt, VOLUME c:\qb45\works

TREATMENT MEAN VARIANCE
----
 1 36.800 4.700
 2 38.667 2.333
 3 46.500 12.333
 (GRAND MEAN = 40.500)

BARTLETT'S TEST (FOR EQUALITY OF VARIANCES)

CHI-SQUARE = 1.482
(DEGREES OF FREEDOM = 2)

ANOVA TABLE:

SOURCE OF SUM OF   MEAN
VARIATION SQUARES D.F. SQUARE   F

TREATMENT 222.533 2 111.267 16.561
ERROR   60.467 9 6.719

TOTAL 283.000 11

TEST OF TREATMENT DIFFERENCES
REJECT HO: AT LEAST SOME OF THE TREATMENTS ARE SIGNIFICANTLY DIFFERENT

SCHEFFE'S MULTIPLE COMPARISONS (ALL PAIRWISE)

COMPARISON Q F ACCEPT/REJECT
----
 1 2 0.986 0.486 A
 1 3 5.579 15.561 R
 2 3 3.957 7.828 R

FOR CONTRAST COEFFICIENTS: 0 -1 1
Q = 3.957 F = 7.828 DECISION: REJECT H^0

P4b: PROGRAM ANOVA
======================
OPTION 2: RANDOMIZED BLOCK DESIGN
 MEASUREMENTS FILE: ex11322.txt, VOLUME c:\qb45\works

TREATMENT MEAN VARIANCE
----
 1 14.500 4.333
 2 16.750 2.250
 3 18.750 4.917
 (GRAND MEAN = 16.667)

BARTLETT'S TEST (FOR EQUALITY OF VARIANCES)

CHI-SQUARE = 0.421
(DEGREES OF FREEDOM = 2)

ANOVA TABLE:

SOURCE OF VARIATION	SUM OF SQUARES	D.F.	MEAN SQUARE	F
TREATMENT	36.167	2	18.083	34.264
BLOCKS	31.333	3	10.444	19.790
ERROR	3.167	6	0.528	

| TOTAL | 70.667 | 11 | | |

TEST OF TREATMENT DIFFERENCES
REJECT HO: AT LEAST SOME OF THE TREATMENTS ARE SIGNIFICANTLY
DIFFERENT

SCHEFFE'S MULTIPLE COMPARISONS (ALL PAIRWISE)

COMPARISON	Q	F	ACCEPT/REJECT
----
1 2	4.380	9.592	R
1 3	8.273	34.225	R
2 3	3.893	7.579	R

FOR CONTRAST COEFFICIENTS: 1 0 -1
Q = 8.273 F = 34.225 DECISION: REJECT H0

P4c: PROGRAM ANOVA
======================
OPTION 3: LATIN SQUARE DESIGN
 MEASUREMENTS FILE: ex11332d.txt, VOLUME
ASSIGNMENTS MAP FILE: ex11332d.txt FROM VOLUME

TREATMENT	MEAN	VARIANCE

1	46.500	195.000
2	47.000	164.667
3	36.500	281.667
4	30.750	83.583
(GRAND MEAN =	40.188)	

BARTLETT'S TEST (FOR EQUALITY OF VARIANCES)

CHI-SQUARE = 0.937
(DEGREES OF FREEDOM = 3)

ANOVA TABLE:

```
SOURCE OF   SUM OF         MEAN
VARIATION   SQUARES  D.F.  SQUARE      F
------------------------------------------------------------
TREATMENT   755.688    3    251.896   12.401
ROWS        1924.688   3    641.563   31.585
COLUMNS     128.188    3     42.729    2.104
ERROR       121.875    6     20.313
------------------------------------------------------------
TOTAL       2930.438  15
------------------------------------------------------------
```

TEST OF TREATMENT DIFFERENCES
REJECT HO: AT LEAST SOME OF THE TREATMENTS ARE SIGNIFICANTLY DIFFERENT

SCHEFFE'S MULTIPLE COMPARISONS (ALL PAIRWISE)

```
------------------------------------------------------------
COMPARISON   Q       F     ACCEPT/REJECT
------------------------------------------------------------
 1 2       0.157   0.008    A
 1 3       3.138   3.282    A
 1 4       4.942   8.142    R
 2 3       3.295   3.618    A
 2 4       5.099   8.667    R
 3 4       1.804   1.085    A
```

FOR CONTRAST COEFFICIENTS: 0 .5 -.5 0
Q =   3.295 F =   3.618 DECISION: ACCEPT HO

3.4.2 Program CANCOR

P1: Canonical correlation analysis is performed. Two data sets are needed. In the example one containing the densities of three species and the other the measurements of two soil variables, both Arrangement 1. Data entry is by variable. The results are stored on a PRINTDA file. The canonical scores are stored on separate files.

P2: The columns are plots and the rows variables

EX1241x.txt Three species.

```
80  70  80  60  55  40  45  30  25  30  20  25  30  35  25  20  25
90  80  80  70  60  60  50  50  60  50  40  45  45  40  55  50  30
10  20  40  50  40  90  60  80  50  60  90  90  80  85  80  70  75
```

Data entered as:
80
70
80

60
55
40
45
30
25
30
.
.
.
90
60
80
50
60
90
90
80
85
80
70
75
85

EX1241Y.txt Two soil variables
7.5 7.5 7.5 7.0 6.5 6.0 6.0 6.5 5.5 6.0 5.5 5.0 5.5 5.0 4.5 5.5 6.0 5.5
45 40 35 45 65 55 45   50 65 60 55   60 55 75 70 55 60 65
Data entered as:
7.5
7.5
7.5
7
6.5
6
6
6.5
5.5
6
5.5
.
.
.
5.5
6
5.5
7.5
7
5.5
6
6.5

P3:
PROGRAM CANCOR

```
======================
```
DATA FILE (LARGER SET): EX1241X.TXT
DATA FILE (SMALLER SET): EX1241Y.TXT
NUMBER OF VARIABLES IN LARGER SET = 3
NUMBER OF VARIABLES IN SMALLER SET = 2
NUMBER OF MEASUREMENTS = 18

MATRIX PRODUCT

```
    0.9251    -0.5879
    0.1874    -0.0292
```

CANONICAL CORRELATIONS

CANONICAL VARIATE   1: R-SQUARED = 0.7907 R = 0.8892 OR 88.3%
(CUMULATIVE: 88.3%)

CANONICAL VARIATE   2: R-SQUARED = 0.1052 R = 0.3243 OR 11.7%
(CUMULATIVE:100.0%)

SIGNIFICANCE TEST FOR CANONICAL CORRELATIONS

WILK'S LAMBDA =   0.1873
BARTLETT'S CHI-SQUARE =   23.452
(DEGREES OF FREEDOM = 6)

VARIABLE WEIGHTS (TRANSFORMATION COEFFICIENTS) - SMALLER SET

VARIATE 1 :
   1.2119   0.2769
VARIATE 2 :
  -1.2137   -1.6927

CANONICAL SCORES - LARGER SET

SET 1
```
    0.4151    0.3837    0.3523    0.2516    0.2137   -0.0126
   -0.0754    0.1195   -0.1133    0.0188   -0.1761   -0.3082
   -0.1761   -0.2139   -0.4088   -0.1761    0.0188   -0.1133
```

SET 2
```
   -0.0769    0.1150    0.3069    0.0868   -0.5172    0.0304
    0.4143    0.0586   -0.1897   -0.1615    0.1942    0.1660
    0.1942   -0.4098   -0.0541    0.1942   -0.1615   -0.1897
```

VARIABLE WEIGHTS (TRANSFORMATION COEFFICIENTS) - LARGER SET

VARIATE 1 :
   0.7669   -0.2229   -0.4870
VARIATE 2 :
  -0.0418   1.8482   1.6613

CANONICAL SCORES - LARGER SET
```
------------------------------------------------------
SET  1
   0.5078   0.4007   0.3978   0.1974   0.2331  -0.1472
   0.0779  -0.1581  -0.0948  -0.0620  -0.2653  -0.2357
  -0.1410  -0.1013  -0.2219  -0.2033  -0.1123  -0.0718

SET  2
   0.0689  -0.0457   0.2772   0.1677  -0.2774   0.5502
  -0.2280   0.1076  -0.0982  -0.2204  -0.0070   0.1323
  -0.0343  -0.0967   0.2519  -0.0514  -0.5393   0.0426
```

STRUCTURE CORRELATIONS, SMALLER SET
```
------------------------------------------------------
CANONICAL VARIATE  1 :
  0.9869  -0.7076

CANONICAL VARIATE  2 :
  0.1615  -0.7066
```

STRUCTURE CORRELATIONS, LARGER SET
```
------------------------------------------------------
CANONICAL VARIATE  1 :
  0.9672   0.8469  -0.9178

CANONICAL VARIATE  2 :
  0.2122   0.4172   0.1432
```

REDUNDANCIES, SMALLER SET
```
------------------------------------------------------
  0.5830   0.0276   (TOTAL =    0.6106)
```

REDUNDANCIES, LARGER SET
```
------------------------------------------------------
 0.658  0.008   (TOTAL =   0.666)
```

U-CANONICAL SCORES (LARGER SET) FILE: ULARGER.TXT
V-CANONICAL SCORES (SMALLER SET) FILE: VSMALLER.TXT

3.4.3 Program CSA

3.4.3.1 The balanced design

This is described as the balanced design (of the dendro-gram) in the main text of Statistical Ecology.

P1: Data entry is from a single Arrangement 1 file, containing two or more score matrices. The data file has specific properties (Example 12.5.1.1):

1. Character states are coded as integer numbers, e.g., simple "+" as 1 and sumbol "-" as 2.
2. The coded values are entered by character.
3. The C/A values are entered in the last vector under the score matrix[2].

The number of score matrices is not limited and the order in which they are arranged is arbitrary. Options permit to establish a firm order for score matrices and characters regardless of the order they are entered. Also, selection of a subset of score matrices is possible by specification entered when requested in the start-up dialogue. The results are stored on files by type. A very detailed printable PRINTDA file and subsidiary files are created. Run programs CSASEL and CSALIMS with files created by the main program.

Program CSASEL prints the product and correlation matrices and creates new file of chord distances, etc. There is option for pooling partial products between hierarchical levels. The names of the input and output files are recorded in the PRINTDA file. The PRINTDA files under the same file name are overwritten in successive runs.
Program CSALIMS prints the results I tabular , including the probabilistic limits determined in a Monte Carlo method for the comparison statistics.

<u>Note</u>: Respond with N to "Draw correlation graphs" .

---

[2]For technical reasons to avoid scrambled limits for relevé correlation on the first variable, the first row in each score matrix may be duplicated to create a dummy variable. This has not been done in the example.

P2a: EX12511.txt Scores of stem succulents in two communities and cover abundance values. Rows are characters. The columns represent character set types.

Community 1

	a	b	c	d	e	f	g
1. Tall	1	2	2	2	1	2	2
2. Green	1	1	1	1	1	2	1
3. Spiny	1	2	1	1	1	1	1
4. Fleshy	2	2	1	2	1	1	1
5. Flat	2	2	1	2	2	2	2
C/A	25	15	7	3	5	1	2

Community 2

	a	b	c	d	e	f	g
1. Tall	1	2	1	2	2	1	2
2. Green	1	2	2	2	1	1	1
3. Spiny	1	2	1	1	2	2	1
4. Fleshy	1	2	2	2	2	2	2
5. Flat	2	2	2	2	2	2	2
C/A	58	5	29	1	1	3	1

Data entered as

1
2
2
2
1
2
2
1
.
.
.
2
2
2
2
25
15
7
3
5
1
2
1
2
1
2
2
1
2
1
2

2
2
.
.
.
2
2
2
2
2
58
5
29
1
1
3
1

Program CSA/1c tasks:
==============
  (i) Character set analysis .
  (ii) Probabilistic limits set in random simulation.
  (iii) Partial correlation graphs drawn.

NUMBER OF CHARACTER SCORE MATRICES: 2
NUMBER OF CHARACETRS: 5
NAME OF SCORE MATRICES AND C/A VECTORS FILE: ex12511.txt VOLUME c:\csar
NUMBER OF ITERATIONS FOR LIMITS: 100
NUMBER OF CHARACTER SET TYPES PER CHARACTER SCORE MATRIX:
CSM 1 : 7
CSM 2 : 7
KEY CODE FOR CSM IDENTITIES:
MATRIX 1 IS CSM # 1
MATRIX 2 IS CSM # 2
NUMBER OF STATES PER CHARACTER:
CHR 2
CHR 2
CHR 2
CHR 2
CHR 2
NEW ORDER OF CHARACTERS
 1 2 3 4 5
NUMBER OF POSSIBLE CSTS PER LEVEL
LEVEL 5 -- 2
LEVEL 4 -- 4
LEVEL 3 -- 8
LEVEL 2 -- 16
LEVEL 1 -- 32
BLOCK SIZE BY LEVEL
LEVEL 5 -- 16
LEVEL 4 -- 8
LEVEL 3 -- 4
LEVEL 2 -- 2

LEVEL 1 -- 1

P4a. X,0 DATA ANALYZED
==================
!!!!!! "Character" in the text refers to character level in the character state hierarchy.!!!!!!

CUMULATIVE C/A ESTIMATES
CM 1
CHARACTER 1
 30 1 / 28 2 / 0 0 / 0 0 / 0 0 / 0 0 / 0 0 /
CHARACTER 2
 30 1 / 27 3 / 1 4 / 0 0 / 0 0 / 0 0 / 0 0 /
CHARACTER 3
 30 1 / 15 6 / 12 5 / 1 7 / 0 0 / 0 0 / 0 0 /
CHARACTER 4
 25 2 / 15 12 / 9 9 / 3 10 / 5 1 / 1 13 / 0 0 /
CHARACTER 5
 25 4 / 15 24 / 7 17 / 3 20 / 5 2 / 1 26 / 2 18 /
CM 2
CHARACTER 1
 90 1 / 8 2 / 0 0 / 0 0 / 0 0 / 0 0 / 0 0 /
CHARACTER 2
 61 1 / 6 4 / 29 2 / 2 3 / 0 0 / 0 0 / 0 0 /
CHARACTER 3
 58 1 / 5 8 / 29 3 / 1 7 / 1 6 / 3 2 / 1 5 /
CHARACTER 4
 58 1 / 5 16 / 29 6 / 1 14 / 1 12 / 3 4 / 1 10 /
CHARACTER 5
 58 2 / 5 32 / 29 12 / 1 28 / 1 24 / 3 8 / 1 20 /
CHARACTER 5 LEVEL 1 DF= 16
NOMINAL PRODUCTS
   832.8750   130.3750
   3941.8750
GLOBAL CORRELATIONS
    1.0000   0.0720
    1.0000
SUMS OF PRODUCT MATRIX (SPECIFIC)
 455 154
 2121
COVARIANCE MATRIX (SPECIFIC)
 28.4375 9.625
 132.5625
CORRELATION MATRIX (SPECIFIC)
 1 .1567634
 1
CHARACTER 4 LEVEL 2 DF= 8
NOMINAL PRODUCTS
   377.8750   -23.6250
   1820.8750
SUMS OF PRODUCT MATRIX (SPECIFIC)
 165.5 -288
 1060.5

COVARIANCE MATRIX (SPECIFIC)
 20.6875 -36
 132.5625
CORRELATION MATRIX (SPECIFIC)
 1 -.6874453
 1
CHARACTER 3 LEVEL 3 DF= 4
NOMINAL PRODUCTS
   212.3750   264.3750
   760.3750
SUMS OF PRODUCT MATRIX (SPECIFIC)
 113.75 205.75
 485.25
COVARIANCE MATRIX (SPECIFIC)
 28.4375 51.4375
 121.3125
CORRELATION MATRIX (SPECIFIC)
 1 .8757524
 1
CHARACTER 2 LEVEL 4 DF= 2
NOMINAL PRODUCTS
   98.6250   58.6250
   275.1250
SUMS OF PRODUCT MATRIX (SPECIFIC)
 98.5 53.5
 65
COVARIANCE MATRIX (SPECIFIC)
 49.25 26.75
 32.5
CORRELATION MATRIX (SPECIFIC)
 1 .6686194
 1
CHARACTER 1 LEVEL 5 DF= 1
NOMINAL PRODUCTS
   0.1250   5.1250
   210.1250
SUMS OF PRODUCT MATRIX (SPECIFIC)
 .125 5.125
 210.125
COVARIANCE MATRIX (SPECIFIC)
 .125 5.125
 210.125
CORRELATION MATRIX (SPECIFIC)
 1 1
 1
M,0 DATA ANALYZED
=================
CUMULATIVE C/A ESTIMATES
CM 1
CHARACTER 1
16.57143 1 / 41.42857 2 / 0 0 / 0 0 / 0 0 / 0 0 / 0 0 /
CHARACTER 2
16.57143 1 / 33.14286 3 / 8.285714 4 / 0 0 / 0 0 / 0 0 / 0 0 /

CHARACTER 3
16.57143 1 / 8.285714 6 / 24.85714 5 / 8.285714 7 / 0 0 / 0 0 / 0 0 /
CHARACTER 4
8.285714 2 / 8.285714 12 / 16.57143 9 / 8.285714 10 / 8.285714 1 / 8.285714 13 / 0
0 /
CHARACTER 5
8.285714 4 / 8.285714 24 / 8.285714 17 / 8.285714 20 / 8.285714 2 / 8.285714 26 /
8.285714 18 /
CM 2
CHARACTER 1
42 1 / 56 2 / 0 0 / 0 0 / 0 0 / 0 0 / 0 0 /
CHARACTER 2
28 1 / 28 4 / 14 2 / 28 3 / 0 0 / 0 0 / 0 0 /
CHARACTER 3
14 1 / 14 8 / 14 3 / 14 7 / 14 6 / 14 2 / 14 5 /
CHARACTER 4
14 1 / 14 16 / 14 6 / 14 14 / 14 12 / 14 4 / 14 10 /
CHARACTER 5
14 2 / 14 32 / 14 12 / 14 28 / 14 24 / 14 8 / 14 20 /
CHARACTER 5 LEVEL 1 DF= 16
NOMINAL PRODUCTS
   375.4464   170.3750
   1071.8750
GLOBAL CORRELATIONS
   1.0000   0.2686
   1.0000
SUMS OF PRODUCT MATRIX (SPECIFIC)
171.6326 174
686
COVARIANCE MATRIX (SPECIFIC)
10.72704 10.875
42.875
CORRELATION MATRIX (SPECIFIC)
1 .5070927
1
CHARACTER 4 LEVEL 2 DF= 8
NOMINAL PRODUCTS
   203.8138   -3.6250
   385.8750
SUMS OF PRODUCT MATRIX (SPECIFIC)
51.48981 -29
343
COVARIANCE MATRIX (SPECIFIC)
6.436226 -3.625
42.875
CORRELATION MATRIX (SPECIFIC)
1 -.2182179
1
CHARACTER 3 LEVEL 3 DF= 4
NOMINAL PRODUCTS
   152.3240   25.3750
    42.8750
SUMS OF PRODUCT MATRIX (SPECIFIC)

```
 77.2347  0
 24.5
COVARIANCE MATRIX (SPECIFIC)
 19.30867  0
 6.125
CORRELATION MATRIX (SPECIFIC)
 1  0
 1
CHARACTER  2  LEVEL  4  DF= 2
NOMINAL PRODUCTS
    75.0893    25.3750
    18.3750
SUMS OF PRODUCT MATRIX (SPECIFIC)
 55.78062  14.5
 12.25
COVARIANCE MATRIX (SPECIFIC)
 27.89031  7.25
 6.125
CORRELATION MATRIX (SPECIFIC)
 1 .5547002
 1
CHARACTER  1  LEVEL  5  DF= 1
NOMINAL PRODUCTS
    19.3087    10.8750
     6.1250
SUMS OF PRODUCT MATRIX (SPECIFIC)
 19.30866  10.875
 6.125
COVARIANCE MATRIX (SPECIFIC)
 19.30866  10.875
 6.125
CORRELATION MATRIX (SPECIFIC)
 1  1
 1
X-M,0 DATA ANALYZED
===================
CUMULATIVE C/A ESTIMATES
CM  1
CHARACTER  1
 13.42857  1 /-13.42857  2 / 0  0 / 0  0 / 0  0 / 0  0 / 0  0 /
CHARACTER  2
 13.42857  1 /-6.142857  3 /-7.285714  4 / 0  0 / 0  0 / 0  0 / 0  0 /
CHARACTER  3
 13.42857  1 / 6.714286  6 /-12.85714  5 /-7.285714  7 / 0  0 / 0  0 / 0  0 /
CHARACTER  4
 16.71428  2 / 6.714286  12 /-7.571428  9 /-5.285714  10 /-3.285714  1 /-7.285714  13 / 0
 0 /
CHARACTER  5
 16.71428  4 / 6.714286  24 /-1.285714  17 /-5.285714  20 /-3.285714  2 /-7.285714  26 /-
 6.285714  18 /
CM  2
CHARACTER  1
 48  1 /-48  2 / 0  0 / 0  0 / 0  0 / 0  0 / 0  0 /
```

CHARACTER 2
33 1 /-22 4 / 15 2 /-26 3 / 0 0 / 0 0 / 0 0 /
CHARACTER 3
44 1 /-9 8 / 15 3 /-13 7 /-13 6 /-11 2 /-13 5 /
CHARACTER 4
44 1 /-9 16 / 15 6 /-13 14 /-13 12 /-11 4 /-13 10 /
CHARACTER 5
44 2 /-9 32 / 15 12 /-13 28 /-13 24 /-11 8 /-13 20 /
CHARACTER 5 LEVEL 1 DF= 16
NOMINAL PRODUCTS
   457.4283   -163.1428
   2870.0000
GLOBAL CORRELATIONS
     1.0000   -0.1424
     1.0000
SUMS OF PRODUCT MATRIX (SPECIFIC)
 220.6325 -81.57142
 1435
COVARIANCE MATRIX (SPECIFIC)
 13.78953 -5.098214
 89.6875
CORRELATION MATRIX (SPECIFIC)
 1 -.1449697
 1
CHARACTER 4 LEVEL 2 DF= 8
NOMINAL PRODUCTS
   236.7958   -81.5714
   1435.0000
SUMS OF PRODUCT MATRIX (SPECIFIC)
 125.8469 -272.9285
 717.5
COVARIANCE MATRIX (SPECIFIC)
 15.73086 -34.11606
 89.6875
CORRELATION MATRIX (SPECIFIC)
 1 -.9082736
 1
CHARACTER 3 LEVEL 3 DF= 4
NOMINAL PRODUCTS
   110.9489   191.3571
   717.5000
SUMS OF PRODUCT MATRIX (SPECIFIC)
 77.05611 95.96424
 408.25
COVARIANCE MATRIX (SPECIFIC)
 19.26403 23.99106
 102.0625
CORRELATION MATRIX (SPECIFIC)
 .9999999 .5410568
 1
CHARACTER 2 LEVEL 4 DF= 2
NOMINAL PRODUCTS
   33.8928   95.3928

309.2500
SUMS OF PRODUCT MATRIX (SPECIFIC)
 11.35203 14.82143
 21.25
COVARIANCE MATRIX (SPECIFIC)
 5.676015 7.410713
 10.625
CORRELATION MATRIX (SPECIFIC)
 1 .9542755
 1
CHARACTER 1 LEVEL 5 DF= 1
NOMINAL PRODUCTS
   22.5408   80.5714
   288.0000
SUMS OF PRODUCT MATRIX (SPECIFIC)
 22.5408 80.5714
 288
COVARIANCE MATRIX (SPECIFIC)
 22.5408 80.5714
 288
CORRELATION MATRIX (SPECIFIC)
 1 .9999999
 1
SCORES

Simulated correlations by level in file GRAPH, volume c:\csar
Specific products in file PRD_SPEC, volume c:\csar
Specific correlations in file COR_SPEC,volume c:\csar
Nominal products in file PRD_NOM, volume c:\csar
Nominal correlations in file COR_NOM,volume c:\csar
Run program CSASORTE to plot limits.
 Run program CSASELEC to select levels for further analysis.

Program CSASEL – tables are printed. Have the following information on hand: number of iterations, number of CST matrices and Number of variables. The input file is cor_spec.txt, cor_nom.txt, prd_nom or prd_spec.

PROGRAM: CSA/SELECTOR 3

Input file name:cor_spec.txt VOLUME c:\csar ARE PRINTED.

LEVEL FIRST NUMBER IN ROW.
AVERAGES LAST ROW AND SECOND LAST COLUMN.
STANDARD DEVIATIONS LAST COLUMN.
GRAND AVERAGE, LAST NUMBER IN LAST ROW.

DATA TYPE (a)
1	0.157	0.156	nd
2	-0.687	-0.688	nd
3	0.876	0.875	nd
4	0.669	0.668	nd
5	1.000	1.000	nd

```
      0.402    0.402    0.000
```

DATA TYPE (b)
```
  1   0.507    0.507     nd
  2  -0.218   -0.219     nd
  3   0.000    0.000     nd
  4   0.555    0.554     nd
  5   1.000    1.000     nd

      0.368    0.368    0.000
```

DATA TYPE (c)
```
  1  -0.145   -0.145     nd
  2  -0.908   -0.909     nd
  3   0.541    0.541     nd
  4   0.954    0.954     nd
  5   1.000    0.999     nd

      0.288    0.288    0.000
```

Program CSALIMS – probabilistic limits are printed for the correlation grap. Have to input when requested by the program the name of the graph matrix (GRAPH.TXT), number of CST matrices, number of variables, and number of iterations.

Program CSA_Lims -
Limits are printed for specific correlation.
===
Data file: c:\csar\GRAPH.TXT
Limit points for specific sorrelation generated
in randomisation in program CSA are printed in
descnding order.
Confidence interval %: 80
Number of iterations: 100
Number of character matrices: 2
Number of variables: 5

DATA TYPE 1
Level 5
COMPARISON 1 2 : .1786323 -.1677943
Level 4
COMPARISON 1 2 : .3567123 -.294719
Level 3
COMPARISON 1 2 : .4183433 -.4439784
Level 2
COMPARISON 1 2 : .7814595 -.8615385
Level 1
COMPARISON 1 2 : .9831905 .7825079

DATA TYPE 2
Level 5
COMPARISON 1 2 : .2010076 -.2

```
Level 4
COMPARISON  1  2 : .4 -.2182179
Level 3
COMPARISON  1  2 : .5070925 -.5025189
Level 2
COMPARISON  1  2 : .8240419 -.8944269
Level 1
COMPARISON  1  2 : .9769  .8541985


DATA TYPE  3
Level 5
COMPARISON  1  2 : .1809891 -.2243212
Level 4
COMPARISON  1  2 : .198148 -.3904833
Level 3
COMPARISON  1  2 : .399897 -.5396967
Level 2
COMPARISON  1  2 : .829072 -.8708838
Level 1
COMPARISON  1  2 : 1 -1
```

3.4.3.2 The unbalanced design

Under development.

3.4.4 Program DISC

P1: Discriminant coefficients and scores are computed for two groups and an external vector. Each group is described by a mean vector and a covariance matrix, in different files. The covariance files are Arrangement 2. The files for the mean vectors and external vector are Arrangement 3. Canonical scores are stored on a file.

```
P2: EX1821M1.txt
16.167
23.5
 3
 EX1821M2.txt
 9.5
36.167
 9
 EX1821X.txt
10
30
10
```

COV1821cov1.txt
19.367
4.1
9.4
47.5
-9.8
8.8
COV1821cov2.txt
4.7
-10.5
-0.6
158.167
-20.46
8

P3: PROGRAM DISC
==================
DISC - discriminant analysis performed on two groups X1,X2
and one external individual X. Program reads upper half
of group covriance matrices COV1,COV2, group mean vectors
M1,M2, and X from folder where the program is located. Data
format: single string,one number per line.

GROUP 1 :
COVARIANCE FILE:ex1821cov1.txt
MEAN VECTOR FILE:ex1821m1.txt
GROUP 2 :
COVARIANCE FILE:ex1821cov2.txt
MEAN VECTOR FILE:ex1821m2.txt
GROUP SIZES ARE: 6 6
EXTERNAL VECTOR FILE:ex1821x.txt
MISCLASSIFICATION COSTS ARE: 1 1
NUMBER OF VARIABLES: 3
Matrix :ex1821cov1.txt
 96.835 20.5 47
 20.5 237.5 -49
 47 -49 44
Vector :ex1821m1.txt
 16.167 23.5 3
Matrix :ex1821cov2.txt
 120.335 -32 44
-32 1028.335 -151.3
 44 -151.3 84
Vector :ex1821m2.txt
 9.5 36.167 9
POOLED COVARIANCE MATRIX:
 12.0335 -3.2 4.4
-3.2 102.8335 -15.13
 4.4 -15.13 8.4

INVERSE OF POOLED COVARIANCE MATRIX:
 .10600718 -6.6274053e-3 -6.7464789e-2
-6.6274053e-3 1.3645088e-2 2.8048901e-2

-6.7464789e-2 2.8048901e-2 .20490773

DISCRIMINANT COEFFICIENTS:
 1.1954879 -.38532065 -2.0345296

EXTERNAL VECTOR X: 10 30 10

1ST GROUP CENTROID: 16.167 23.5 3
2ND GROUP CENTROID: 9.5 36.167 9

STATISTICS:
EXTERNAL VECTOR X TO 1ST GROUP CENTROID DISTANCE: 24.118865
EXTERNAL VECTOR X TO 2ND GROUP CENTROID DISTANCE: .93948684
DISCRIMINANT SCORE Y OF EXTERNAL VECTOR X(zero origin in centre on line Y):
-11.589689
RECOMMENDED ASSIGNMENT BASED ON DISTANCES:GROUP 2
ASSIGNMENT CONSIDERING CANONICAL SCORE, GROUP SIZE AND
MISCLASSIFICATION COST:
Empirical proportions are used!
$u = 0$
SINCE Y < uASSIGNMENT TO GROUP 2 IS INDICATED

3.4.5 Program EMVO

P1: The hypothesis that given group mean vectors have common expectation is tested under different options. Several groups (k) and several variables (s) are permitted. The relation $s<k<n_{min}$ must hold true. In this n_{min} is the size of the smallest group. Unequal replicate numbers are permitted. Users specify a critical Theta-value (from Heck's charts) at selected probability (alpha) and constants (s, m, n) determined by the program. The options include:
1. Multivariate analysis of variance (EMVO).
2. Canonical group analysis (CANON).
3. Profile analysis (PROFILE).
Canonical scores (Option 2) are stored on disk file under user specified name. All other information are stored in the PRINTDA file. Data entry is by relevé-by-groups (Arrangement 1), all groups stored within same work file.

P2: ex1151.txt -- Three peat bogs are involved (A,B,C) and cover estimates for 3 species within 15 randomly sited plots (replicates).

A   B   C

Plot	Species X_1	X_2	Species X_1	Species X_2	X_1	X_2
1	16	11	18	27	43	14
2	15	12	19	34	38	17
3	18	9	23	32	37	18
4	12	17	19	36	39	12
5	18	15	25	31	44	14
6	9	9	24	29	40	13
7	15	17	22	35	42	11
8	9	16	27	30	40	13
9	16	18	25	34	36	14
10	18	9	19	32	39	12
11	17	16	18	32	39	17
12	13	12	26	35	45	16
13	16	9	25	31	40	14
14	15	17	26	34	40	11
15	14	10	26	34	38	12

Data set presented as:

16
11
15
12
18
9
12
17
18
15
9
9
15
17
9
16
16
18
18
.
.
.
.
39
12
39
17
45
16
40
14
40
11
38
12

P4a: PROGRAM: EMVO
=====================
OPTION 1 SELECTED - EMVO (COMPARE K MEAN VECTORS)
DATA FILE:ex1151.txt
NUMBER OF VARIABLES: 2
NUMBER OF GROUPS: 3
GROUP SIZES: 15 , 15 , 15 ,
MEAN VECTOR, GROUP 1
  14.733 13.133
MEAN VECTOR, GROUP 2
  22.800 32.400
MEAN VECTOR, GROUP 3
  40.000 13.867
MEAN VECTOR, ALL GROUPS:
  25.844 19.800
MEAN VECTOR, GROUPS
  13.933 27.600 26.933
GRAND MEAN = 22.822222
H MATRIX
   4996.57778-   724.13333
- 724.13333   3576.13333
E MATRIX
   363.33333-   13.26667
-   13.26667   331.06667
INVERSE OF MATRIX E (INV(E))
     .00276   .00011
     .00011   .00302
H*INV(E) MATRIX
    13.69222-   1.63859
-   1.60095   10.73770
EIGENVALUE 1 =   14.40710 (CUMULATIVE: 14.40710)
EIGENVECTOR: .91310-   .40774
EIGENVALUE 2 =   10.02281 (CUMULATIVE: 24.42992)
EIGENVECTOR: .39979   .91661
TEST FOR EQUALITY OF THE 3 MEAN VECTORS
THETA =   .935 (S = 2.0, M = - .5 AND N = 19.5)
THETA (AT CHOSEN alpha) = .185
REJECT HO. MEAN VECTORS DO NOT HAVE A COMMON EXPECTATION.
SIMULTANEOUS LIMITS (ALL PAIRWISE, FOR EACH VARIABLE):

VARIABLE	GROUPS		LOWER/UPPER LIMITS	
1	1	2-	11.38278-	4.75056
1	1	3-	28.58278-	21.95056
1	2	3-	20.51611-	13.88389
2	1	2-	22.43211-	16.10123
2	1	3-	3.89877	2.43211
2	2	3	15.36789	21.69877

CONTRAST COEFFICIENTS FOR MEAN VECTORS: 1 -1 0
DESIGN VECTOR FOR VARIABLES: 1 1
LOWER LIMIT = -   31.82928   UPPER LIMIT = -   22.83738
CONTRAST COEFFICIENTS FOR MEAN VECTORS: 1 -.5 -.5
DESIGN VECTOR FOR VARIABLES: 1 0

LOWER LIMIT = -   19.53850   UPPER LIMIT = -   13.79483

P4b: PROGRAM: EMVO
=====================
OPTION 2 SELECTED - CANONICAL GROUPS ANALYSIS
DATA FILE:ex1151.txt
NUMBER OF VARIABLES: 2
NUMBER OF GROUPS: 3
GROUP SIZES: 15 , 15 , 15 ,
MEAN VECTOR, GROUP 1
 14.733 13.133
MEAN VECTOR, GROUP 2
 22.800 32.400
MEAN VECTOR, GROUP 3
 40.000 13.867
MEAN VECTOR, ALL GROUPS:
 25.844 19.800
MEAN VECTOR, GROUPS
 13.933 27.600 26.933
GRAND MEAN = 22.822222
H MATRIX
   4996.57778-   724.13333
-   724.13333   3576.13333
E MATRIX
    363.33333-   13.26667
-    13.26667   331.06667

INVERSE OF MATRIX E (INV(E))
      .00276   .00011
      .00011   .00302
H*INV(E) MATRIX
    13.69222-   1.63859
-    1.60095   10.73770

EIGENVALUE 1 =   14.40710 (CUMULATIVE: 14.40710)
EIGENVECTOR: .91310-   .40774
EIGENVALUE 2 =   10.02281 (CUMULATIVE: 24.42992)
EIGENVECTOR: .39979   .91661
TEST FOR EQUALITY OF THE 3 MEAN VECTORS
THETA =   .935 (S = 2.0, M = - .5 AND N = 19.5)
THETA (AT CHOSEN alpha) = .185
REJECT HO. MEAN VECTORS DO NOT HAVE A COMMON EXPECTATION.
SIMULTANEOUS LIMITS (ALL PAIRWISE, FOR EACH VARIABLE):

VARIABLE	GROUPS		LOWER/UPPER LIMITS	
1	1	2-	11.38278-	4.75056
1	1	3-	28.58278-	21.95056
1	2	3-	20.51611-	13.88389
2	1	2-	22.43211-	16.10123
2	1	3-	3.89877	2.43211
2	2	3	15.36789	21.69877

CONTRAST COEFFICIENTS FOR MEAN VECTORS: 1 -1 0
DESIGN VECTOR FOR VARIABLES: 1 1
LOWER LIMIT = -   31.82928   UPPER LIMIT = -   22.83738
CONTRAST COEFFICIENTS FOR MEAN VECTORS: 1 -.5 -.5

DESIGN VECTOR FOR VARIABLES: 1 0
LOWER LIMIT = -   19.53850   UPPER LIMIT = -   13.79483

CANONICAL GROUP ANALYIS:
CANONICAL (DISCRIMINANT) SCORES OF INDIVIDUALS, SET 1

- 5.40086-	6.72170-	2.75919-	11.49969-	5.20562
- 10.97708-	8.76039-	13.83125-	8.25503-	2.75919
- 6.52645-	8.54790-	4.58539-	8.76039-	6.81933
- 10.09847-	12.03953-	7.57166-	12.85500-	5.33772
- 5.43535-	9.70797-	3.10379-	6.56093-	11.22405
- 12.13715-	6.05557-	5.33772-	5.64783-	5.64783
18.02960	12.24089	10.92006	15.19268	18.94270
15.69804	18.33971	15.69804	11.63791	15.19268
13.15399	19.04032	15.29030	16.51352	14.27958

GROUP CENTROID SCORES:
- 7.427- 7.917 15.345

CANONICAL (DISCRIMINANT) SCORES OF INDIVIDUALS, SET 2

- 12.00187-	11.48506-	13.03549-	8.10141-	7.53586
- 16.63363-	6.90203-	10.21739-	5.58563-	13.03549
- 7.01905-	12.28464-	13.83508-	6.90203-	13.71806
3.46340	10.27944	10.04540	12.11265	9.92838
7.69537	12.39542	9.81136	12.67819	8.44623
8.04643	13.99459	9.92838	13.07799	13.07799
1.54236	2.29321	2.81002-	1.89002-	1.94215
- .57362-	1.60725-	.57362-	1.25619-	1.89002
2.69300	4.17516	.34298-	2.40684-	2.28982

GROUP CENTROID SCORES:
- 10.553 10.332   .221
DISCRIMINANT SCORES IN FILE scores.txt

P4c: PROGRAM: EMVO
======================
OPTION 3 SELECTED - PROFIL ANALYSIS
DATA FILE:ex1151.txt
NUMBER OF VARIABLES: 2
NUMBER OF GROUPS: 3
GROUP SIZES: 15 , 15 , 15 ,
MEAN VECTOR, GROUP 1
 14.733 13.133
MEAN VECTOR, GROUP 2
 22.800 32.400
MEAN VECTOR, GROUP 3
 40.000 13.867
MEAN VECTOR, ALL GROUPS:
 25.844 19.800
MEAN VECTOR, GROUPS
 13.933 27.600 26.933
GRAND MEAN = 22.822222
H MATRIX
  10020.97778- 4300.26667
- 4300.26667 3576.13333
E MATRIX
  720.93333- 344.33333

-   344.33333   331.06667

INVERSE OF MATRIX E (INV(E))
    .00276   .00287
    .00287   .00600
H*INV(E) MATRIX
   15.29317   2.91688
-    1.60095   9.13674

EIGENVALUE 1 =   14.40716 (CUMULATIVE: 14.40716)
EIGENVECTOR: .87494   .48422
EIGENVALUE 2 =   10.02276 (CUMULATIVE: 24.42992)
EIGENVECTOR: -   .29057-   .95685

TEST FOR EQUALITY OF THE 3 PROFILES
THETA =   .935 (S =   1.0, M =   .0 AND N = 20.0)
THETA (AT CHOSEN alpha) = .185
REJECT HO: PROFILES ARE NOT PARALLEL
TEST FOR THE EQUALITY OF GROUP EFFECTS
SUM OF SQUARES AMONG =   7124.44444
SUM OF SQUARES WITHIN =   667.86667
F =   224.017
(NUMERATOR DEGREES OF FREEDOM: 2, DENOMINATOR DEGREES OF FREEDOM: 42)

3.4.6 Program FANOVA

P1: Factorial Model I ANOVA is performed. Complete randomised and randomised block designs are available as options. The program can handle up to 4 factors. The same number of replicates is assumed in each factor combination. The replicates are entered by factor combination (Arrangement 1). For example, given 2 factors A, B with 2 and 3 levels respectively, the treatment combinations are:
(A1B1)
(A1B2)
(A1B3)
(A2B1)
(A2B2)
(A2B3)

P2: ex11342.txt Four replicates within 6 treatment combinations (Example 11.3.4.2):

$$A_1B_1 \qquad A_1B_2 \qquad A_1B_3 \qquad A_2B_1 \qquad A_2B_2 \qquad A_2B_3$$

	40	32	32	35	38	44
Replicates	36	34	31	33	40	42
	38	35	33	32	42	44
	38	35	36	36	40	42
Mean	38	34	33	34	40	43

The data are presented as:

40
36
38
38
32
34
35
35
32
31
33
36
35
33
32
36
38
40
42
40
44
42
44
42

P3: PROGRAM FANOVA
=======================
INPUT (RAW DATA) FILE: ex11342, VOLUME c:\qb45\works
NUMBER OF FACTORS: 2
NUMBER OF TREATMENTS: 6
NUMBER OF REPLICATES PER TREATMENT COMBINATION: 4
OPTION: COMPLETE RANDOMIZED DESIGN.
BARTLETT'S TEST (EQUALITY OF EXPECTED VARIANCES)

CHI-SQUARE = 1.212
(DEGREES OF FREEDOM = 5
MEANS FOR MAIN EFFECTS AND FIRST ORDER INTERACTIONS

LEVEL   MEAN

-> GRAND MEAN = 37.000
FACTOR A
 1   35.000
 2   39.000
FACTOR B
 1   36.000

```
2  37.000
3  38.000
ANOVA TABLE
```

```
SOURCE OF SUM OF   MEAN
VARIATION SQUARES D.F. SQUARE   F
```

```
TREATMENT
 A  96.000  1  96.000  34.560
 B  16.000  2  8.000   2.880
 AB 208.000 2 104.000 37.440
ERROR 50.000 18 2.778
```

```
TOTAL  370.000 23
```

```
SCHEFFE'S MULTIPLE COMPARISONS (ALL MAIN EFFECTS, PAIRWISE)
```

```
COMPARISON  Q  F
```

```
=== FACTOR 1
(NUMERATOR DEGREES OF FREEDOM: 1 )
(DENOMINATOR DEGREES OF FREEDOM: 18 )
 1 2 5.879 34.560
=== FACTOR 2
(NUMERATOR DEGREES OF FREEDOM: 2 )
(DENOMINATOR DEGREES OF FREEDOM: 18 )
 1 2 1.200 0.720
 1 3 2.400 2.880
 2 3 1.200 0.720
```

3.4.7 Program LIMITS

P1: The equality of the expectation of a sample mean vector and a standard vector is tested. Two input data files are required:
1. Arrangement 1 file for sample data entered by variables.
2. Arrangement 3 file for standard vector.
Confidence limits are determined for the variables. User must be able to specify a critical F-value for the degrees of freedom determined by the program.

P2: Example 10.4.1. Chemical analysis of soil samples from a field yielded the following data for 3 varoab;es and 10 replicates:

Sampling unit	1	2	3	4	...	7	8	9	10	Total
Nitrate N	7.5	10.9	10.8	12.2	...	3.0	6.6	12.9	4.0	79.2
Ammoniacal N	20.1	20.4	17.9	18.0	...	21.0	16.1	13.8	18.0	172.8

Elementary K   42.0   43.2   40.4   84.3 ... 110.4 101.4 87.7 115.2   793.9

Ex1041n.txt
7.5
10.9
10.8
12.2
5.1
6.2
3
6.6
12.9
4
20.1
20.4
17.9
18
15.4
12.1
21
16.1
13.8
18
42
43.2
40.4
84.3
69.5
99.8
110.4
101.4
87.7
115.2

EX 1041X.txt
7 .92
17 .28
79 .39

P3: PROGRAM LIMITS
========================
DATA FILE: EX1041N.TXT, VOLUME:
NUMBER OF VARIABLES: 3
NUMBER OF OBSERVATIONS PER VARIABLE: 10
POPULATION MEANS FILE: EX1041X.TXT
SAMPLE MEANS:

  7.9200 17.2800 79.3900
SAMPLE COVARIANCE MATRIX

  12.5218 -1.0784 -52.3609
  -1.0784   8.5351 -26.6524
 -52.3609 -26.6524 840.4343
POPULATION MEAN VECTOR

```
-----------------------------------------------------
   7.0000   0.9200   17.0000
```

INVERSE OF SAMPLE COVARIANCE MATRIX:
```
-----------------------------------------------------
   0.120714   0.042995   0.008884
   0.042995   0.145355   0.007288
   0.008884   0.007288   0.001975
```
TEST FOR EQUALITY OF MEAN VECTORS:

```
-----------------------------------------------------
```
GENERALIZED DISTANCE =   7.9928
T =   25.2754
F = 165.6267
H0 REJECTED. SAMPLE AND POP'N MEAN VECTORS DO NOT HAVE SAME EXPECTATION.
CONFIDENCE LIMITS FOR THE SAMPLE MEANS:

```
---------------------------------------------------------
```
VARIABLE	CONTRAST	ERROR	LOWER	UPPER LIMITS
```
---------------------------------------------------------
```
1	0.9200	4.5836	-3.6636	5.5036
2	16.3600	3.7843	12.5757	20.1443
3	62.3900	37.5517	24.8383	99.9417

3.4.8 Program TESTCOV1

P1: The hypothesis that a given standard square symmetric matrix is the expectation of a given sample covariance matrix is tested. Two data files are required, the sample covariance and the standard matrix, both Arrangement 2 files. Users need to know the sample size based on which the sample covariance matrix is computed.

P2: cov1041.txt
12.5218
-1.0784
-52.3609
8.5351
-26.6524
840.4343
    cov1051p.txt
10.3014
-1.36403
-60.3333
9.17830
-22.4160

830.445

P3: PROGRAM TESTCOV1
=====================
POPULATION COVARIANCE MATRIX FILE: covp1051.txt, VOLUME
SAMPLE COVARIANCE MATRIX FILE: cov1041.txt, VOLUME
NUMBER OF VARIABLES (ORDER OF cov1041.txt): 3
SAMPLE SIZE: 10
POPULATION COVARIANCE MATRIX (V)

```
  10.3014  -1.3640  -60.3333
  -1.3640   9.1783  -22.4160
 -60.3333 -22.4160  830.4450
```
INVERSE OF POPULATION COVARIANCE MATRIX (INV(V))

```
  0.2052   0.0716   0.0168
  0.0716   0.1416   0.0090
  0.0168   0.0090   0.0027
```

SAMPLE COVARIANCE MATRIX (S)

```
  12.5218  -1.0784  -52.3609
  -1.0784   8.5351  -26.6524
 -52.3609 -26.6524  840.4343
```

TEST FOR EQUALITY OF POP'N AND SAMPLE COVARIANCE MATRICES

DETERMINANT OF SAMPLE COVARIANCE MATRIX (S) = 53538.73047
DETERMINANT OF POP'N COVARIANCE MATRIX (V) = 34697.19141
TRACE OF MATRIX PRODUCT S*(INV(V)):   3.62415
CHI-SQUARE =   1.71360
CORRECTED CHI-SQUARE (FOR SMALL SAMPLE SIZES) =   1.507
(DEGREES OF FREEDOM = 6)

3.4.9 Program TESTCOV2

P1: The equality of the expectations of two or more sample covariance matrices is tested. Data entry is by relevé-by-group (see Section 3.4.5). The file type is Arrangement 1.

P2: EX1151.txt
16
11
15
12
18
9
12
17
18

15
9
9
.
.
.
39
12
39
17
45
16
40
14
40
11
38
12

P3:PROGRAM TESTCOV/2
==========================
DATA FILE: EX1151.TXT, VOLUME
NUMBER OF VARIABLES: 2
NUMBER OBSERVATIONS PER VARIABLE: 45
NUMBER OF GROUPS: 3
GROUP SIZES: 15 15 15
COVARIANCE MATRIX, GROUP 1

   8.495 -0.962
  -0.962 12.410
INVERSE OF COVARIANCE MATRIX, GROUP 1

   0.119   0.009
   0.009   0.081
DETERMINANT 1 = 104.4966

COVARIANCE MATRIX, GROUP 2

  11.029   0.657
   0.657   6.257
INVERSE OF COVARIANCE MATRIX, GROUP 2

   0.091 -0.010
  -0.010   0.161
DETERMINANT 2 = 68.57552
COVARIANCE MATRIX, GROUP 3

   6.429 -0.643
  -0.643   4.981
INVERSE OF COVARIANCE MATRIX, GROUP 3

   0.158   0.020
   0.020   0.203

```
DETERMINANT  3  =  31.60714
POOLED COVARIANCE MATRIX
-----------------------------------------------------------
    8.651   -0.316
   -0.316    7.883
INVERSE OF POOLED COVARIANCE MATRIX
-----------------------------------------------------------
   0.116   0.005
   0.005   0.127
DETERMINANT  4  =  68.09045
TEST FOR EQUALITY OF THE  3  COVARIANCE MATRICES
-----------------------------------------------------------
CHI-SQUARE =    4.329
(B =  4.648522  AND 1/C =  .931217 )
DEGREES OF FREEDOM =  6
```

3.5 Importance *a posteriori*

3.5.1 Program WEIGHINF

P1: Variables are given weights based on logarithmic expressions of entropy and information. The exact form of the weight function depends on the option selected. The data elements represent diagnostic categories coded as numerals. The data entry is by variable and storage is in an Arrangement 1 file. Sample size (the number of observations per variable) is assumed to be very large.

P2: Example 14.4.1. Sequence of arrival for the same five species on 80 different occasions:

Species	Arrival sequences			
A	2442132214	5222221334	2322412211	3212522224
	3255342231	3452124415	2442225221	2241434544
B	1533451125	3333554253	3214141142	232111333
	5322251352	5534435522	3313313112	3414551215
C	5321325553	2151113522	1533533335	5553231152
	2544533125	2323313354	1251152335	1353325453
D	3214213332	1515332411	5455225523	4135354511
	1133425513	1215252233	5125531553	5522212322
E	4155544441	4444445145	4141354454	1444445445
	4411114444	4141541141	4534444444	4135143131

The data are presented as:

```
ex1441.txt
2
4
4
2
1
```

```
3
2
2
1
4
5
2
2
2
2
1
3
.
.
.
3
5
1
4
3
1
3
1
```

P3:
PROGRAM WEIGHINF
==============================
DATA FILE: ex1441.txt
NUMBER OF VARIABLES: 5
NUMBER OF OBJECTS: 80
WEIGHTING OPTION: INTERACTION INFORMATION
VARIABLE 1

 STATES: 2 4 1 3 5
FREQUENCY: 33 17 12 10 8
VARIABLE 2

 STATES: 1 5 3 4 2
FREQUENCY: 18 16 23 9 14
VARIABLE 3

 STATES: 5 3 2 1 4
FREQUENCY: 23 25 14 14 4
VARIABLE 4

 STATES: 3 2 1 4 5
FREQUENCY: 17 19 17 6 21
VARIABLE 5

 STATES: 4 1 5 3
FREQUENCY: 44 19 12 5

JOINT FREQUENCIES

VECTOR 1 : 2 1 5 3 4 FREQUENCY = 4
VECTOR 2 : 4 5 3 2 1 FREQUENCY = 8
VECTOR 3 : 4 3 2 1 5 FREQUENCY = 4
VECTOR 4 : 2 3 1 4 5 FREQUENCY = 2
VECTOR 5 : 1 4 3 2 5 FREQUENCY = 6
VECTOR 6 : 3 5 2 1 4 FREQUENCY = 6
VECTOR 7 : 1 2 5 3 4 FREQUENCY = 6
VECTOR 8 : 5 3 2 1 4 FREQUENCY = 3
VECTOR 9 : 2 3 1 5 4 FREQUENCY = 10
VECTOR 10 : 2 3 5 1 4 FREQUENCY = 4
VECTOR 11 : 2 5 1 3 4 FREQUENCY = 2
VECTOR 12 : 3 2 5 4 1 FREQUENCY = 4
VECTOR 13 : 2 1 3 5 4 FREQUENCY = 8
VECTOR 14 : 2 4 3 5 1 FREQUENCY = 3
VECTOR 15 : 4 1 5 2 3 FREQUENCY = 5
VECTOR 16 : 5 1 2 3 4 FREQUENCY = 1
VECTOR 17 : 5 2 4 3 1 FREQUENCY = 4
SET CONTAINS 5 VARIABLE(S). INFORMATION QUANTITIES:

----
 1. JOINT INFORMATION = 409.047
 2. MUTUAL INFORMATION = 358.378
 3. EQUIVOCATION INFORMATION = 50.670
 4. MULTIPLE OF ENTROPY = 575.254
 5. MULTIPLE OF JOINT ENTROPY = 216.876
VARIABLE 2 HAS RANK 1 (INTERACTION INFORMATION = 125.336)
JOINT FREQUENCIES

VECTOR 1 : 2 5 3 4 FREQUENCY = 4
VECTOR 2 : 4 3 2 1 FREQUENCY = 8
VECTOR 3 : 4 2 1 5 FREQUENCY = 4
VECTOR 4 : 2 1 4 5 FREQUENCY = 2
VECTOR 5 : 1 3 2 5 FREQUENCY = 6
VECTOR 6 : 3 2 1 4 FREQUENCY = 6
VECTOR 7 : 1 5 3 4 FREQUENCY = 6
VECTOR 8 : 5 2 1 4 FREQUENCY = 3
VECTOR 9 : 2 1 5 4 FREQUENCY = 10
VECTOR 10 : 2 5 1 4 FREQUENCY = 4
VECTOR 11 : 2 1 3 4 FREQUENCY = 2
VECTOR 12 : 3 5 4 1 FREQUENCY = 4
VECTOR 13 : 2 3 5 4 FREQUENCY = 8
VECTOR 14 : 2 3 5 1 FREQUENCY = 3
VECTOR 15 : 4 5 2 3 FREQUENCY = 5
VECTOR 16 : 5 2 3 4 FREQUENCY = 1
VECTOR 17 : 5 4 3 1 FREQUENCY = 4
SET CONTAINS 4 VARIABLE(S). INFORMATION QUANTITIES:

----
 1. JOINT INFORMATION = 280.292
 2. MUTUAL INFORMATION = 233.042
 3. EQUIVOCATION INFORMATION = 47.251
 4. MULTIPLE OF ENTROPY = 449.918
 5. MULTIPLE OF JOINT ENTROPY = 216.876

VARIABLE 4 HAS RANK 2 (INTERACTION INFORMATION = 110.402)
JOINT FREQUENCIES

VECTOR 1 : 2 5 4 FREQUENCY = 8
VECTOR 2 : 4 3 1 FREQUENCY = 8
VECTOR 3 : 4 2 5 FREQUENCY = 4
VECTOR 4 : 2 1 5 FREQUENCY = 2
VECTOR 5 : 1 3 5 FREQUENCY = 6
VECTOR 6 : 3 2 4 FREQUENCY = 6
VECTOR 7 : 1 5 4 FREQUENCY = 6
VECTOR 8 : 5 2 4 FREQUENCY = 4
VECTOR 9 : 2 1 4 FREQUENCY = 12
VECTOR 10 : 3 5 1 FREQUENCY = 4
VECTOR 11 : 2 3 4 FREQUENCY = 8
VECTOR 12 : 2 3 1 FREQUENCY = 3
VECTOR 13 : 4 5 3 FREQUENCY = 5
VECTOR 14 : 5 4 1 FREQUENCY = 4
SET CONTAINS 3 VARIABLE(S). INFORMATION QUANTITIES:
----
 1. JOINT INFORMATION = 164.739
 2. MUTUAL INFORMATION = 122.640
 3. EQUIVOCATION INFORMATION = 42.099
 4. MULTIPLE OF ENTROPY = 326.315
 5. MULTIPLE OF JOINT ENTROPY = 203.675
VARIABLE 1 HAS RANK 3 (INTERACTION INFORMATION = 94.796)
JOINT FREQUENCIES

VECTOR 1 : 5 4 FREQUENCY = 14
VECTOR 2 : 3 1 FREQUENCY = 11
VECTOR 3 : 2 5 FREQUENCY = 4
VECTOR 4 : 1 5 FREQUENCY = 2
VECTOR 5 : 3 5 FREQUENCY = 6
VECTOR 6 : 2 4 FREQUENCY = 10
VECTOR 7 : 1 4 FREQUENCY = 12
VECTOR 8 : 5 1 FREQUENCY = 4
VECTOR 9 : 3 4 FREQUENCY = 8
VECTOR 10 : 5 3 FREQUENCY = 5
VECTOR 11 : 4 1 FREQUENCY = 4
SET CONTAINS 2 VARIABLE(S). INFORMATION QUANTITIES:
----
 1. JOINT INFORMATION = 58.720
 2. MUTUAL INFORMATION = 27.844
 3. EQUIVOCATION INFORMATION = 30.876
 4. MULTIPLE OF ENTROPY = 208.782
 5. MULTIPLE OF JOINT ENTROPY = 180.939
VARIABLE 3 HAS RANK 4 (INTERACTION INFORMATION = 27.844)
JOINT FREQUENCIES

VECTOR 1 : 4 FREQUENCY = 44
VECTOR 2 : 1 FREQUENCY = 19
VECTOR 3 : 5 FREQUENCY = 12
VECTOR 4 : 3 FREQUENCY = 5
SET CONTAINS 1 VARIABLE(S). INFORMATION QUANTITIES:

----
1. JOINT INFORMATION = 20.656
2. MUTUAL INFORMATION = 0.000
3. EQUIVOCATION INFORMATION = 20.656
4. MULTIPLE OF ENTROPY = 90.247
5. MULTIPLE OF JOINT ENTROPY = 90.247
VARIABLE 5 HAS RANK 5

3.5.2 Program WEIGHSCP

P1: Variables are given weights according to sum of squares, cross products, multiple correlation, or specific variance depending on the option chosen. Input is from an Arrangement 2 disk file of the covariance or cross product matrix.

P2: cov4221cov1.txt (created in program CORRELATION from data in Table 4.1.2.2.1. Note: 16.857x7/6=19.6667)

19.66667
13.83333
-1.499998
16
-5.166667
5.571427

P3a: PROGRAM WEIGHSCP
============================
PRODUCTS FILE: cov1411, VOLUME c:\qb45\works
PRODUCTS FILE: COV41221.TXT, VOLUME
WEIGHTS FILE: WEIGHTSC.TXT, VOLUME
NUMBER OF VARIABLES: 3
WEIGHTING OPTION: COMMON VARIANCE
PRODUCTS MATRIX

    19.667   13.833   -1.500
    13.833   16.000   -5.167
    -1.500   -5.167   5.571
SUMMARY
VARIABLE RANK   COMMON VARIANCE
----
    1   1   14.216
    2   2   3.098
    3   3   0.000
P3b: PROGRAM WEIGHSCP
============================
PRODUCTS FILE: COV41221.TXT, VOLUME
WEIGHTS FILE: WEIGTS2.TXT, VOLUME
NUMBER OF VARIABLES: 3

WEIGHTING OPTION: SPECIFIC VARIANCE
PRODUCTS MATRIX

```
    19.667     13.833      -1.500
    13.833     16.000      -5.167
    -1.500     -5.167       5.571
```

SUMMARY
VARIABLE RANK   SPECIFIC VARIANCE

```
    1    1       5.451
    2    2       3.172
    3    3       2.761
```

P3c: PROGRAM WEIGHSCP
==========================
PRODUCTS FILE: COV41221.TXT, VOLUME
WEIGHTS FILE: WEIGHTSM.TXT, VOLUME
NUMBER OF VARIABLES: 3
WEIGHTING OPTION: MULTIPLE CORRELATION
PRODUCTS MATRIX

```
    19.667     13.833      -1.500
    13.833     16.000      -5.167
    -1.500     -5.167       5.571
```

SUMMARY
VARIABLE RANK   MULTIPLE CORRELATION

```
    1    2       0.293
    2    1       0.802
    3    3       0.000
```

P3d: PROGRAM WEIGHSCP
==========================
PRODUCTS FILE: COV41221.TXT, VOLUME
WEIGHTS FILE: WEIGHTSS.TXT, VOLUME
NUMBER OF VARIABLES: 3
WEIGHTING OPTION: SUM OF SQUARES
PRODUCTS MATRIX

```
    19.667     13.833      -1.500
    13.833     16.000      -5.167
    -1.500     -5.167       5.571
```

SUMMARY
VARIABLE RANK   SUM OF SQUARES

```
    1    2       8.849
    2    1      29.628
    3    3       2.761
```

3.6 Cluster recognition

3.6.1 Program ALC

P1: Average Link Clustering is performed on a similarity matrix. Similarity values are expected in the 0 and 1 range. The data file is Arrangement 2. The cluster descriptors are stored on file for input in program TREE.

P2: Example 17.2.1 Table:

	Sensitivity class					
	1	2	3	4	5	6
Low	0	50	82	59	13	0
Medium	100	29	13	16	22	0
High	0	21	5	25	65	100

The data are presented as cos1721.txt:
1
.4715603
.1562979
.2422574
.3149936
0
1
.8957855
.9698004
.6176686
.3414747
.9999999
.9413289
.2886834
6.011457E-02
.9999999
.5948676
.3785273
.9999999
.930663
1
P3:
PROGRAM ALC
====================
SIMILARITIES FILE: COS1721.TXT, VOLUME
NUMBER OF INDIVIDUALS: 6
CLUSTERING PASS # 1

NUMBER OF OBJECTS IN FUSION GROUP: 2
AVERAGE SIMILARITY: 0.970
OBJECTS:
  2 4
CLUSTERING PASS # 2

NUMBER OF OBJECTS IN FUSION GROUP: 2
AVERAGE SIMILARITY: 0.931
OBJECTS:
  5 6

CLUSTERING PASS # 3

NUMBER OF OBJECTS IN FUSION GROUP: 3
AVERAGE SIMILARITY: 0.926
OBJECTS:
  2   4   3
CLUSTERING PASS # 4

NUMBER OF OBJECTS IN FUSION GROUP: 5
AVERAGE SIMILARITY: 0.419
OBJECTS:
  2   4   3   5   6
CLUSTERING PASS # 5

NUMBER OF OBJECTS IN FUSION GROUP: 6
AVERAGE SIMILARITY: 0.396
OBJECTS:
  1   2   4   3   5   6
DENDROGRAM DATA STORED IN FILE DENRalc.TXT

3.6.2 Program SLC

P1:   Single Link Clustering is performed on distances or similarity values input from an Arrangement 2 file. Dendrogram data are stored in the TREE file.

P2: dist1711.txt (generated from data in ex41221.txt:

```
0
 5.830952
 6.557438
 7.681146
 4.24264
 13.78405
 15.55635
 0
 1
 2.236068
 6.480741
 9.273619
 11.22497
 0
 2
 7.28011
 9.433981
 11.35782
 0
 8.544004
```

8.062258
10.63015
0
12.72792
13.34166
0
4.472136
0

P3: PROGRAM SLC
====================
SIMILARITY (OR DISTANCE) FILE:dist1711.txt
DENDROGRAM DATA IN FILE: tree.txt
ORDER OF DENDROGRAM: 7
INPUT MATRIX OPTION: DISTANCES
CLUSTERING PASS 1

NUMBER OF INDIVIDUALS IN FUSION GROUP: 2
DISTANCE AT FUSION: 1.000
INDIVIDUALS: 2 3

CLUSTERING PASS 2

NUMBER OF INDIVIDUALS IN FUSION GROUP: 3
DISTANCE AT FUSION: 2.000
INDIVIDUALS: 2 3 4

CLUSTERING PASS 3

NUMBER OF INDIVIDUALS IN FUSION GROUP: 2
DISTANCE AT FUSION: 4.243
INDIVIDUALS: 1 5

CLUSTERING PASS 4

NUMBER OF INDIVIDUALS IN FUSION GROUP: 2
DISTANCE AT FUSION: 4.472
INDIVIDUALS: 6 7

CLUSTERING PASS 5

NUMBER OF INDIVIDUALS IN FUSION GROUP: 5
DISTANCE AT FUSION: 5.831
INDIVIDUALS: 2 3 4 1 5

CLUSTERING PASS 6

NUMBER OF INDIVIDUALS IN FUSION GROUP: 7
DISTANCE AT FUSION: 8.062
INDIVIDUALS: 2 3 4 1 5 6 7
Dendrogram data in file tree.txt

3.6.3 Program SSA

P1:   Sum of squares clustering is performed on a distance matrix stored in an Arrangement 2 file. Dendrogram data are stored in the TREE file.

P2: DIST1711 (see listing in Section 3.6.2 above)

P3: PROGRAM SSA
====================
DISTANCES FILE: dist1711.txt
NUMBER OF OBJECTS: 7
CLUSTERING PASS 1

NUMBER OF INDIVIDUALS IN FUSION GROUP: 2
FUSION SUM OF SQUARES: 0.500
INDIVIDUALS: 2 3
CLUSTERING PASS 2

NUMBER OF INDIVIDUALS IN FUSION GROUP: 3
FUSION SUM OF SQUARES: 3.333
INDIVIDUALS: 2 3 4
CLUSTERING PASS 3

NUMBER OF INDIVIDUALS IN FUSION GROUP: 2
FUSION SUM OF SQUARES: 9.000
INDIVIDUALS: 1 5
CLUSTERING PASS 4

NUMBER OF INDIVIDUALS IN FUSION GROUP: 2
FUSION SUM OF SQUARES: 10.000
INDIVIDUALS: 6 7
CLUSTERING PASS 5

NUMBER OF INDIVIDUALS IN FUSION GROUP: 5
FUSION SUM OF SQUARES: 66.400
INDIVIDUALS: 1 5 2 3 4
CLUSTERING PASS 6

NUMBER OF INDIVIDUALS IN FUSION GROUP: 7
FUSION SUM OF SQUARES: 247.429
INDIVIDUALS: 1 5 2 3 4 6 7
DENDROGRAM DATA STORED IN FILE tree1731.txt

3.6.4 Program TRGRPS

P1: A deterministic answer is supplied in response to the proposition that a collection naturally divides into R groups along discontinuities. If R groups cannot be found, the quest continues for R-1 groups, R-2 groups, until a specified condition is satisfied. The input data contains distance values from an Arrangement 2 file. Dendrogram data are stored on file.

P2: Input is DISTRAW4, a distance matrix whose elements are computed by METRICS from the 2 x 85 raw data matrix below. X_1 *and* X_2 *are co-ordinate sets:*

X_1	195	255	175	232	123	192	114	170	226	165
	212	221	278	257	319	334	373	332	368	380
	403	421	468	431	406	465	479	478	526	540
	531	632	682	626	615	681	652	656	695	733
	728	668	627	689	610	600	641	576	552	505
	588	488	469	429	361	295	355	575	381	412
	365	404	430	460	451	462	571	615	625	675
	611	683	729	805	830	791	855	844	891	931
	911	932	859	862	926					

X_2	125	178	195	222	238	258	317	310	330	380
	392	448	421	491	481	534	507	583	555	595
	579	541	584	621	638	682	641	614	609	628
	679	668	655	612	595	597	569	521	529	574
	487	491	443	418	410	372	347	320	328	299
	269	258	189	209	142	129	208	641	342	379
	378	414	400	390	430	449	96	74	120	156
	179	213	261	260	270	331	345	401	380	440
	491	152	103	164	176					

195
255
175
232
123
192
114
170
226
165
212
221
278
257
319
334
373
332
368
380
403
421
468
.
.

380
440
491
152
103
164
176
P3: PROGRAM TRGRPS
DISTANCEs FILE: DISTRAW4.TXT
ORDER OF DISTANCE MATRIX: 85
MINIMUM GROUP SIZE EXPECTED: 4
NUMBER OF GROUPS TO BE FORMED: 4
INITIAL NEIGBOURHOOD RADIUS: 20.24846
RADIUS INCREMENT: 3.59
 4 GROUPS FORMED
MAXIMUM NEIGHBOURHOOD RADIUS= 77.68845
GROUP 1
 1 2 4 6 8 7
 9 56 3 10 11 12
 14 13 15 16 19 20
 21 22 17 18 23 27
 28 26 24 25 29 30
 58 31 32 33 34 35
 36 37 38 42 39 41
 40 43 45 46 47 48
 49 51 50 52 44 5
 55 57 53 54
GROUP 2
 59 61 60 63 62 64
 65 66
GROUP 3
 67 68 69 70 72 71
 73 74 75 76 77 79
 78 80 81
GROUP 4
 82 85 83 84
ELAPSED TIME= 419.7969 SECONDS
P5: Edited plot of the data listed under P3. Data and graph adapted from Orlóci (1978).

3.7 Trend seeking

3.7.1 Programs CONA

P1: A complete canonical contingency table analysis is performed by options CONA (root program) and RECIPROCAL
ORDERING if chosen.   CONA offers two weighting options:

pre-analysis data adjustments by autocorrelation analysis and covariance analysis with position, root transformation, and control over the size and scaling of graphs. Should the row or column categories of the contingency table have un-equal sizes, select the "adjust the raw contingency scores to block sizes" option. Raw scores and block sizes are input from disk file. These files have Arrangement 1. Data entry is by row. Separate files are created for Eigenvalues, canonical scores, and many other properties. These are identified in the PRINTDA file. The user should be familiar with the method, file names and data dimensions before running the program.

P2a: Raw data Example 17.5.2.1:

Quadrat group		a	b	c	Total
Species	A	2	15	71	88
group	B	74	21	2	97
	C	16	112	35	163
Total		92	148	108	348

and

$$N = \begin{bmatrix} 10\times14 & 10\times20 & 10\times11 \\ 9\times14 & 9\times20 & 9\times11 \\ 7\times14 & 7\times20 & 7\times11 \end{bmatrix} = \begin{bmatrix} 140 & 200 & 110 \\ 126 & 180 & 99 \\ 98 & 140 & 77 \end{bmatrix}$$

The data are presented as:

ex17521f.txt
2
15
71
74
21
2
16
112
35
P2b: Block sizes ex17521n.txt
140
200
110
126
180
99
98

140
77
P3: Program CONA

Input file for raw contingency table data:
C:\Users\Koa\Desktop\EPIC Program Appendix\WORKS CONA\Ex17521f.txt
Number of rows: 3
Number of columns: 3
Program ADJUST:::
Input file for block sizes:
C:\Users\Koa\Desktop\EPIC Program Appendix\WORKS CONA\Ex17521n.txt
Adjusted frequencies stored in file: ADJUSTDAT.txt
Program CONA:::
Input data file: ADJUSTDAT.txt
Weighting option selected: # 3
1. High weight for rows with high total
2. High weight for rows with low total
3. NO WEIGHTING -- Canonical contingency table analysis proper
4. Correspondence analysis
Row totals
    88.882   87.604   171.514
Column totals
    92.525   119.963   135.512
 3 x 3 table
    1.728   9.073   78.081
   71.046   14.113   2.444
   19.750   96.777   54.987
GRAND TOTAL =   348.000
 3 x 3 SCALAR PRODUCT MATRIX
    .2588   -.2105   -.0359
   -.2105   .3905   -.1275
   -.0359   -.1275   .1170
CANONICAL CORRELATIONS AND CHI-SQUARED PARTITIONS
Can corr R 1 =   .7490 F..R^2 =   195.2211 Cum % = 73.2050
Can corr R 2 =   .4531 F..R^2 =   71.4562 Cum % = 100.0000
TEST FOR COMPOSITIONAL SHARPNESS OF THE 9 BLOCKS
Chi squared =   266.6773
Degrees of freedom = 4
Rank = 2
EIGEN-ADJUSTED ROW SCORES
SET 1
    -.4104   .6180   -.1030
SET 2
    -.3442   -.1489   .2544
EIGEN-ADJUSTED COLUMN SCORES
SET 1
    .6401   -.0596   -.3843
SET 2
    -.1503   .3658   -.2212
ROW SCORES (unadjusted)
SET 1
   -1.0801   1.6264   -.2710
SET 2

89

```
        -1.3224      -.5721       .9775
COLUMN SCORES (unadjusted)
SET 1
        1.5632      -.1455      -.9385
SET 2
        -.5635      1.3710      -.8290
```
Row totals in file: rowtots.txt
Column totals in file: coltots.txt
Row canonical scores in file(s): ROWS.txt and rowsadj.txt
 (2 sets of 3 numbers)
Column canonical scores in file(s): COLS.txt and colsadj.txt
 (2 sets of 3 numbers)
Run PLOT with ROWS ... or COLS ... to draw scattergram
Deviations are in file:Dev partitions..txt
Eigenvalues stored in file: EIG.txt
Deviations information stored in file: LAT.txt
Program DEVIATIONS:::
Input data files are: LAT.txt and EIG.txt
The number of canonical variates is: 2
DEVIATIONS FROM RANDOM EXPECTATION (Fhj - Fh.F.j/F..)
=== Table of deviations partitions 1
Table row 1
```
       -29.8833       3.6059      26.2774
```
Table row 2
```
        44.3521      -5.3518     -39.0004
```
Table row 3
```
       -14.4688       1.7459      12.7229
```
=== Table of deviations partitions 2
Table row 1
```
         7.9798     -25.1727      17.1929
```
Table row 2
```
         3.4026     -10.7338       7.3312
```
Table row 3
```
       -11.3824      35.9064     -24.5241
```
=== Sum of deviations partitions tables
Table row 1
```
       -21.9036     -21.5668      43.4703
```
Table row 2
```
        47.7547     -16.0855     -31.6692
```
Table row 3
```
       -25.8512      37.6523     -11.8011
```
Graphs in file: GRAPHS.txt
Program ANGLES AND DISTANCES:::
Input data file:colsadj.txt
Output in files:ANGLES.txt and DISTANCES.txt
Number of axes: 2
Number of units: 3
Print in following order:
Point B; squared distances AB,AC,BC; angle at B
 2:    .755958   1.054450   .449999   82.537306
Program LATGRAPHS:::
Input data file: GRAPHS.txt
Number of deviations partitions: 2

DEVIATIONS PROFILES: rows are plotted
ANGLES PROFILE: columns are the points
DISTANCES PROFILES: columns are the points
The main printable file is: PRINTDACONA.txt
Profiles are stored in KEP files with extension: ..txt
Open and print the PRINTDA file from EDIT
Open and edit KEP files from a paint program
Press any key to exit
Elapsed time: 106.175 seconds

3.7.2 Program EIGENAN

P1: Eigenvalues and Eigenvectors are computed for a square symmetric matrix. The data file is an Arrangement 2. If the input products matrix is a dual matrix, the elements in the Eigenvectors are component scores, identical to those computed in PCAR from the primary data.

P2: cov41221a (computed from ex41221 as a dual matrix in METRICS):

8.06	1.75	1.30	-0.03	5.85	-7.46	-9.46
	1.11	1.32	0.99	0.37	-2.27	-3.27
		1.70	1.37	-0.25	-2.22	-3.22
			1.70	-1.91	-0.22	-1.89
				6.63	-5.84	-4.84
					8.68	9.35
						13.35

The data are input as:
8.0612245
 1.7517007
 1.2993197
-3.4013605e-2
 5.8469388
-7.462585
-9.462585
 1.1088435
 1.3231293
 .98979592
 .3707483
-2.2721088
-3.2721088
 1.7040816
 1.3707483
-.24829932
-2.2244898
-3.2244898
 1.7040816
-1.914966

-.2244898
-1.8911565
 6.6326531
-5.8435374
-4.8435374
 8.6802721
 9.3469388
13.346939

P3: PROGRAM: EIGENAN
==================
INPUT FILE:ex41221a.TXT
ORDER OF QCOV.TXT = 7
MATRIX
 8.06 1.75 1.30 -0.03 5.85 -7.46 -9.46
 1.75 1.11 1.32 0.99 0.37 -2.27 -3.27
 1.30 1.32 1.70 1.37 -0.25 -2.22 -3.22
-0.03 0.99 1.37 1.70 -1.91 -0.22 -1.89
 5.85 0.37 -0.25 -1.91 6.63 -5.84 -4.84
-7.46 -2.27 -2.22 -0.22 -5.84 8.68 9.35
-9.46 -3.27 -3.22 -1.89 -4.84 9.35 13.35
EIGEN UNDERWAY

EIGENVALUE 1 = 32.55746 OR 78.94996 % (CUM. 78.94996 %)
VECTOR:
 2.7577 0.7788 0.7104 0.1696 1.9040 -2.8344 -3.4861
EIGENVALUE 2 = 7.235593 OR 17.54589 % (CUM. 96.49586 %)
VECTOR:
-0.3935 0.6653 0.9713 1.2943 -1.7275 0.1898 -0.9996
EIGENVALUE 3 = 1.445051 OR 3.504164 % (CUM. 100 %)
VECTOR:
-0.5489 0.2445 0.5058 -0.0137 0.1523 -0.7811 0.4411
EIGENVALUES IN FILE EIGEN.TXT, VOLUME
SCORES IN FILE SCORES.TXT
RUN PROGRAM plot OR stereo FOR SCATTERGRAM.
JOB REQUIRED 32.01953 SECONDS

3.7.3 Program MDSCAL

P1: Metric co-ordinates are computed from proximity data. Two data files are involved: one holds the external sample distance matrix (Arrangement 2) and the other optionally an Arrangement 1 file of the co-ordinates of a fixed a priori configuration (entered by axes). If the a priori configuration option is not chosen, the starting configuration is an arbitrary set of random numbers generated in the program. The user must specify the dimensionality (t) of the final configuration. This should be an integer number one higher than the desired dimensionality. The final internal configuration co-ordinates are stored in the PRINTDA file and also stored in an Arrangement 1 file readable by other programs.

P2: A distance matrix based on the mountain transect data in Example 15.1.1:

Elevation	L							
Plot	1	2	3	4	5	6	7	8
Species	40	45	50	47	41	35	25	20
	10	16	22	37	40	45	42	40
	0	2	5	11	13	17	25	31
	0	0	1	1	1	2	5	8

	13	14	15	16	17	18	19	H 20
...	5	2	2	2	0	0	0	0
...	16	13	7	3	2	1	0	0
...	51	40	35	28	25	20	14	10
...	35	52	60	66	60	49	36	30

He data are presented as DIST1511.txt:

```
0
.1052357
.1928823
.4535775
.5617208
.7016984
.8663269
.9703931
1.093145
1.167334
1.23586
1.288905
1.314349
1.359323
1.376889
1.387685
1.408929
1.41097
1.414214
1.414214
0
8.976305E-02
.3505577
.4600158
.6024256
.7734289
.8839278
```

1.017757
1.103469
1.18073
1.241149
1.272548
1.328019
1.355381
1.373655
1.395436
1.398477
1.40345
1.404827
0
.2650412
.3752905
.5183064
.6902354
.8036758
.9441739
1.037839
1.121315
1.186656
1.221721
1.285459
1.320897
1.345767
1.368075
1.371946
1.378391
1.381223
0
.1117995
.2592246
.4531273
.5870704
.7608944
.8887394
.9942795
1.077279
1.126906
1.214774
1.27449
1.317117
1.339107
1.345336
1.355827
1.361776
0
.148774
.3544704
.4975883
.6853689
.8275096

.941213
1.030739
1.086411
1.184333
1.253994
1.30394
1.325353
1.332492
1.344565
1.35194
0
.2226565
.3754752
.5789792
.7384385
.8609736
.9572355
1.019548
1.130211
1.212767
1.272578
1.293022
1.301311
1.315423
1.324997
0
.1562084
.3687551
.5420728
.67285
.7771661
.8531979
.99682
1.100434
1.17919
1.198788
1.208556
1.225642
1.241408
0
.2163448
.3977671
.5318751
.6393189
.7239748
.8920065
1.009244
1.100809
1.119598
1.130184
1.149041
1.169421
0

.1884145
.3241448
.4367113
.5422629
.7568088
.8908262
.9995593
1.016829
1.028012
1.048574
1.076153
0
.1390738
.2635376
.4053572
.6709281
.8122479
.9320757
.9479586
.9588335
.9796052
1.013482
0
.13444
.3074578
.6026279
.7453263
.8708073
.8848024
.8952753
.9158746
.9537796
0
.1907899
.4993093
.6411927
.7704132
.7829524
.7931867
.8137634
.8545639
0
.3129739
.4577488
.5913516
.6041424
.6148981
.6364836
.6785688
0
.1578237
.2943826
.3071375

.3195475
.3435366
.3838668
0
.1392593
.1526568
.1645418
.1881919
.2303963
0
3.066039E-02
3.859274E-02
5.868157E-02
.0939675
0
1.391408E-02
3.894841E-02
7.922188E-02
0
2.516951E-02
6.841571E-02
0
4.913581E-02
0

P3: PROGRAM: MDSCAL
=========================
EXTERNAL DISTANCES FILE: 0
ORDER OF DISTANCE MATRIX: 20
DIMENSIONALITY OF FINAL SOLUTION: 3
STARTING CONFIGURATION OPTION: 2 --Random configuration
Random seed: .58376
INTERNAL DISTANCE OPTION: 1 --Euclidean
ITERATION 1

STRESS VALUE = 36.333771%
GRADIENT LENGTH: 0.347411
(OLD STRESS)-(LAST STRESS) = 6.744191%
ITERATION 2

STRESS VALUE = 29.589579%
GRADIENT LENGTH: 0.201572
(OLD STRESS)-(LAST STRESS) = 1.593375%
ITERATION 3

STRESS VALUE = 27.996202%
GRADIENT LENGTH: 0.078185
(OLD STRESS)-(LAST STRESS) = 0.499091%
ITERATION 4

STRESS VALUE = 27.497112%
GRADIENT LENGTH: 0.060756
(OLD STRESS)-(LAST STRESS) = 0.402439%

ITERATION 5

STRESS VALUE = 27.094673%
GRADIENT LENGTH: 0.103884
(OLD STRESS)-(LAST STRESS) = 0.849578%
ITERATION 6

STRESS VALUE = 26.245096%
GRADIENT LENGTH: 0.121797
(OLD STRESS)-(LAST STRESS) = 3.712784%
ITERATION 7

STRESS VALUE = 22.532312%
GRADIENT LENGTH: 0.451018
(OLD STRESS)-(LAST STRESS) = 6.041601%
ITERATION 8

STRESS VALUE = 16.490711%
GRADIENT LENGTH: 0.346979
(OLD STRESS)-(LAST STRESS) = 3.477190%
ITERATION 9

STRESS VALUE = 13.013521%
GRADIENT LENGTH: 0.536770
(OLD STRESS)-(LAST STRESS) = 3.802992%
ITERATION 10

STRESS VALUE = 9.210528%
GRADIENT LENGTH: 0.319669
(OLD STRESS)-(LAST STRESS) = 2.036245%
ITERATION 11

STRESS VALUE = 7.174283%
GRADIENT LENGTH: 0.499496
(OLD STRESS)-(LAST STRESS) = 1.597909%
ITERATION 12

STRESS VALUE = 5.576375%
GRADIENT LENGTH: 0.304920
(OLD STRESS)-(LAST STRESS) = 0.854067%
ITERATION 13

STRESS VALUE = 4.722308%
GRADIENT LENGTH: 0.436948
(OLD STRESS)-(LAST STRESS) = 0.784837%
ITERATION 14

STRESS VALUE = 3.937470%
GRADIENT LENGTH: 0.236364
(OLD STRESS)-(LAST STRESS) = 0.718650%
ITERATION 15

STRESS VALUE = 3.218821%

GRADIENT LENGTH: 0.468309
(OLD STRESS)-(LAST STRESS) = 0.523250%
ITERATION 16

STRESS VALUE = 2.695571%
GRADIENT LENGTH: 0.206461
(OLD STRESS)-(LAST STRESS) = 0.202114%
ITERATION 17

STRESS VALUE = 2.493457%
GRADIENT LENGTH: 0.205572
(OLD STRESS)-(LAST STRESS) = 0.162143%
ITERATION 18

STRESS VALUE = 2.331314%
GRADIENT LENGTH: 0.227226
(OLD STRESS)-(LAST STRESS) = 0.114450%
ITERATION 19

STRESS VALUE = 2.216864%
GRADIENT LENGTH: 0.138591
(OLD STRESS)-(LAST STRESS) = 0.294180%
ITERATION 20

STRESS VALUE = 1.922685%
GRADIENT LENGTH: 0.221464
(OLD STRESS)-(LAST STRESS) = 0.074666%
ITERATION 21

STRESS VALUE = 1.848019%
GRADIENT LENGTH: 0.114575
(OLD STRESS)-(LAST STRESS) = 0.040933%
ITERATION 22

STRESS VALUE = 1.807086%
GRADIENT LENGTH: 0.147486
(OLD STRESS)-(LAST STRESS) = 0.036017%
ITERATION 23

STRESS VALUE = 1.771068%
GRADIENT LENGTH: 0.081668
(OLD STRESS)-(LAST STRESS) = 0.167999%
ITERATION 24

STRESS VALUE = 1.603069%
GRADIENT LENGTH: 0.191050
(OLD STRESS)-(LAST STRESS) = 0.043851%
ITERATION 25

STRESS VALUE = 1.559218%
GRADIENT LENGTH: 0.061743
(OLD STRESS)-(LAST STRESS) = 0.039954%
ITERATION 26

```
-------------------------------------------------------
STRESS VALUE =  1.519263%
GRADIENT LENGTH:  0.181556
(OLD STRESS)-(LAST STRESS) =  0.044082%
ITERATION  27
-------------------------------------------------------
STRESS VALUE =  1.475181%
GRADIENT LENGTH:  0.064596
(OLD STRESS)-(LAST STRESS) =  0.015382%
ITERATION  28
-------------------------------------------------------
STRESS VALUE =  1.459800%
GRADIENT LENGTH:  0.066939
(OLD STRESS)-(LAST STRESS) =  0.013900%
ITERATION  29
-------------------------------------------------------
STRESS VALUE =  1.445900%
GRADIENT LENGTH:  0.071857
(OLD STRESS)-(LAST STRESS) =  0.012042%
ITERATION  30
-------------------------------------------------------
STRESS VALUE =  1.433858%
GRADIENT LENGTH:  0.078011
(OLD STRESS)-(LAST STRESS) =  0.010880%
THE MAXIMUM NUMBER OF ITERATIONS HAS BEEN REACHED! LAST STRESS =
1.422977%
FINAL CONFIGURATION
-------------------------------------------------------
AXIS  1
  -0.1554 -0.1415 -0.0884 -0.1335 -0.1247 -0.0700
  -0.0835 -0.0690 -0.0050 -0.0260 -0.0479 -0.0238
   0.0493  0.0827  0.1051  0.1532  0.1253  0.1439
   0.1656  0.1436
AXIS  2
  -0.3153 -0.3014 -0.2788 -0.1945 -0.1898 -0.1717
  -0.0925 -0.0720 -0.0387  0.0268  0.0672  0.0821
   0.0991  0.1268  0.1708  0.1820  0.2245  0.2214
   0.2195  0.2345
AXIS  3
  -0.1171 -0.0635 -0.0786 -0.0858 -0.0296 -0.0528
  -0.0534 -0.0014 -0.0033 -0.0375 -0.0051  0.0417
  -0.0047  0.0721  0.0624  0.0796  0.0620  0.0548
   0.0608  0.0994
FINAL CONFIGURATION SCORES STORED IN FILE sco1511.txt
Elapsed time= 88.47852  seconds
```

3.7.4 Program PCAR

P1: Principal Components Analysis is performed on data entered by variable on an Arrangement 1 file. Component coefficients are computed for variables and component scores for relevés. The latter are stored in a separate Arrangement 1 file by components.

P2: ex41221.txt (given in Section 3.1.1).

P3: PROGRAM PCA COORDINATE FILE C:\Users\Koa\Desktop\EPIC Program Appendix\WORKS
====================================
 PCAR\ex41221.txt
NUMBER OF ROWS (VARIABLES): 3
NUMBER OF COLUMNS (INDIVIDUALS): 7
OPTION: VARIANCE,COVARIANCE
VARIANCE,COVARIANCE MATRIX

```
-----------------------------------------------------
    19.667    13.833-    1.500
    13.833    16.000-    5.167
 -  1.500-     5.167     5.571
```
SUBROUTINE EIGENEIGENVECTORS ADJUSTED TO UNIT LENGTH.
 EIGENVALUE

```
-----------------------------------------------------
```
EIGENVALUE 1 =   32.5575 OR 78.95% (CUMULATIVE 78.95%)
EIGENVECTOR: .730 .662- .167
EIGENVALUE 2 =   7.2356 OR 17.55% (CUMULATIVE 96.50%)
EIGENVECTOR: .531- .396 .750
EIGENVALUE 3 =   1.4451 OR   3.50% (CUMULATIVE 100.00%)
EIGENVECTOR:- .430 .636 .640
COMPONENT SCORES:

```
-----------------------------------------------------
```
SET 1 :
 - 2.758-   .779-   .710-   .170-   1.904   2.834
    3.486
SET 2 :
    .394-   .665-   .971-   1.294   1.728-   .190
   1.000
SET 3 :
    .549-   .244-   .506   .014-   .152   .781
 - .441
COMPONENT SCORES STORED IN TEXT FILE sco41221.txt

3.7.5 Program REGRESSION

P1: Regression analysis is performed on one or more factor variables (X) and one response variable (Y). Option is available for polynomial regression. Two input files are required: the X file (Arrangement 1) and the Y file (Arrangement 3). Data entry in the X file is by variable. The user chooses the rejection probability (alpha) and corresponding F value for the degrees of freedom displayed on the screen by the program.

P2: Example 13.2.1

Temperature X_1	Light level X_2	Photosynthetic rate Y
$^{\circ}C$	$nE\ cm^{-2}sec^{-1}$	$mg\ CO_2\ cm^{-2}\ hr^{-1}$
5	5	1
7	30	6
8	10	2
10	10	3
10	30	10
8	5	2

Data are presented as ex1321x.txt for X_1, X_2:

5
7
8
10
10
8
5
30
10
10
30
5

ex1321Y for Y

1
6
2
3
10
2

P3a: PROGRAM: REGRESSION
============================
FILE X (INDEPENDENT VARIABLES): ex1321x.txt
FILE Y (DEPENDENT VARIABLE): ex1321y.txt
OPTION SELECTED: 1
NUMBER OF OBSERVATIONS: 6
T VALUE AT DF = 4 IS 2.776
ANOVA TABLE

Source of variation, sum of squares, DF, mean square, F

REGRESSION   16.05556   1   16.05556   1.53113

```
RESIDUAL    41.94444   4   10.48611
TOTAL       58.00000   5
```

COEFFICIENT OF DETERMINATION = 0.27682
REGRESSION COEFFICIENTS

B(0) = -3.55556 B(1) = 0.94444

Y	Y-HAT	RESIDUAL
1.00000	1.16667	0.16667
6.00000	3.05556	2.94444
2.00000	4.00000	2.00000
3.00000	5.88889	2.88889
10.00000	5.88889	4.11111
2.00000	4.00000	2.00000

LOWER AND UPPER CONFIDENCE LIMITS FOR B-VALUES

B(0)
B(1)   -1.17436   3.06325
LOWER AND UPPER CONFIDENCE LIMITS, Y-HAT AND Y VALUES

Lower and upper limits		Y-hat	Y
-6.17308	8.50641	-10.60516	12.60516
-1.18205	7.29316	-3.93806	15.93806
0.33013	7.66987	-7.70957	11.70957
0.28307	11.49471	-7.59401	13.59401
0.28307	11.49471	-0.59401	20.59401
0.33013	7.66987	-7.70957	11.70957

P3b: PROGRAM: REGRESSION
===========================
FILE X (INDEPENDENT VARIABLES): ex1321x.txt
FILE Y (DEPENDENT VARIABLE): ex1321y.txt
OPTION SELECTED: 2
NUMBER OF OBSERVATIONS: 6
T VALUE AT DF = 3 IS 3.182

X VARIABLE* MEAN   VARIANCE

1	8.00000	3.60000
2	15.00000	140.00000

*IF OP=3 THEN X1=X AND X2=X^2
SSCP MATRIX- CORRECTED FOR MEANS

```
18.00000   35.00000
35.00000  700.00000
```
INVERSE OF SSCP

```
0.06154   -0.00308
-0.00308    0.00158
```
ANOVA TABLE

```
--------------------------------------------------------
Source of variation, sum of squares, DF, mean square, F
--------------------------------------------------------
REGRESSION   52.58901   2   26.29451   14.57839
RESIDUAL      5.41099   3    1.80366
TOTAL        58.00000   5
--------------------------------------------------------
COEFFICIENT OF DETERMINATION =  0.90671
REGRESSION COEFFICIENTS
--------------------------------------------------------
     VALUE    STANDARDIZED   T       VIF
--------------------------------------------------------
B(0)   -3.42198
B( 1 )   0.47692    0.26569   1.43152    1.10769
B( 2 )   0.24044    0.83530   4.50057    1.10769
--------------------------------------------------------
    Y       Y-HAT     RESIDUAL
--------------------------------------------------------
   1.00000    0.16484    0.83516
   6.00000    7.12967    1.12967
   2.00000    2.79780    0.79780
   3.00000    3.75165    0.75165
  10.00000    8.56044    1.43956
   2.00000    1.59560    0.40440
```
LOWER AND UPPER CONFIDENCE LIMITS FOR B-VALUES
```
--------------------------------------------------------
B(0)
B( 1 )   -0.58319    1.53703
B( 2 )    0.07044    0.41044
```
LOWER AND UPPER CONFIDENCE LIMITS, Y-HAT AND Y VALUES
```
--------------------------------------------------------
   Lower and upper limits   Y-hat       Y
--------------------------------------------------------
  -3.39558    3.72525   -4.56227    6.56227
   3.61463   10.64471    0.46667   11.53333
   0.85714    4.73847   -2.69345    6.69345
   0.68810    6.81520   -2.25810    8.25810
   5.29398   11.82689    4.62115   15.37885
  -0.84029    4.03150   -2.91893    6.91893
```

P3c: PROGRAM: REGRESSION
```
==============================
```
FILE X (INDEPENDENT VARIABLES): ex1321x.txt
FILE Y (DEPENDENT VARIABLE): ex1321y.txt
OPTION SELECTED: 3
NUMBER OF OBSERVATIONS: 6
T VALUE AT DF = 3 IS 3.182
```
--------------------------------------------------------
X VARIABLE*  MEAN    VARIANCE
--------------------------------------------------------
   1    8.00000    3.60000
   2   67.00000   856.79999
```

*IF OP=3 THEN X1=X AND X2=X^2
SSCP MATRIX- CORRECTED FOR MEANS

```
   18.00000   276.00000
  276.00000  4284.00000
```
INVERSE OF SSCP

```
    4.57692    -0.29487
   -0.29487     0.01923
```
ANOVA TABLE

Source of variation, sum of squares, DF, mean square, F

```
REGRESSION   16.82681   2    8.41341    0.61303
RESIDUAL     41.17319   3   13.72440
TOTAL        58.00000   5
```

COEFFICIENT OF DETERMINATION = 0.29012

REGRESSION COEFFICIENTS

	VALUE	STANDARDIZED	T	VIF
B(0)	3.22439			
B(1)	-0.92308	-0.51423	-0.11647	82.38448
B(2)	0.12179	1.04674	0.23707	82.38448

Y	Y-HAT	RESIDUAL
1.00000	1.65386	0.65386
6.00000	2.73078	3.26922
2.00000	3.63462	1.63462
3.00000	6.17306	3.17306
10.00000	6.17306	3.82694
2.00000	3.63462	1.63462

LOWER AND UPPER CONFIDENCE LIMITS FOR B-VALUES

```
B(0)
B( 1 )   -26.14239    24.29623
B( 2 )    -1.51293     1.75652
```

LOWER AND UPPER CONFIDENCE LIMITS, Y-HAT AND Y VALUES

Lower and upper limits		Y-hat	Y
-9.98218	13.28991	-15.56378	17.56378
-4.33204	9.79360	-7.74208	19.74207
-3.23642	10.50565	-11.64450	15.64450
-2.10882	14.45495	-11.40662	17.40662
-2.10882	14.45495	-4.40662	24.40662
-3.23642	10.50565	-11.64450	15.64450

3.8 Accessory programs

3.8.1 Program ADJUST

P1: Contingency table entries are adjusted to equal block size. The adjusted data are written on disk file to be read by CONA. The data file is Arrangement 1 and data entry is from two disk files: raw contingency table entries and the block sizes. Entry is by row or column in both, which ever has the lesser dimension.

P1. Ex1721f.txt and ex121n.txt under Section 3.7.1.

P2. PROGRAM: ADJUST
========================
CONTINGENCY TABLE FILE:ex17521f.txt
BLOCK SIZE FILE:ex17521n.txt
NUMBER OF ROWS: 3
NUMBER OF COLUMNS: 3
ADJUSTED OCCUPANCY COUNTS:

1.728	9.073	78.081
71.046	14.113	2.444
19.750	96.777	54.987

ADJUSTED OCCUPANCY COUNTS STORED IN FILE adjdat.txt

3.8.2 Program PLOT

P1: Scattergram is plotted and the graph is stored on file. The file name starts with KEP and is BMP type. Data entry is by axis from an Arrangement 1 co-ordinate file. User specifies the axes to be plotted. The horizontal axis must have the smaller numerical label. The co-ordinates and vital run information are shown in the PRINTDA file. An option is available to generate prime number series for co-ordinates on first plotted axis.

P2: sco1511.txt (created in MDSCAL, Section 3.7.3)
-.1554109
-.1415034
-8.840343E-02
-.1335071

-.1246622
-6.999867E-02
-8.353358E-02
-6.897741E-02
-4.953756E-03
-2.597786E-02
-4.789472E-02
-2.383612E-02
4.933594E-02
8.267437E-02
.1051333
.1531858
.1253051
.143874
.1655907
.1435599
-.3153382
-.3014272
-.2787694
-.1944942
-.1897967
-.1716889
-9.253273E-02
-7.200998E-02
-3.865875E-02
2.682161E-02
6.717698E-02
8.212166E-02
9.906022E-02
.1267897
.1708207
.1819873
.2245268
.2213659
.2195169
.2345283
-.1171344
-6.349274E-02
-7.864727E-02
-8.582487E-02
-2.958582E-02
-.0527752
-5.341254E-02
-1.398055E-03
-3.250984E-03
-3.753535E-02
-5.051509E-03
4.171196E-02
-4.711939E-03
7.213666E-02
6.236947E-02
7.958278E-02
6.198987E-02

5.480028E-02
6.083023E-02
9.939945E-02

P3: PROGRAM PLOT
======================
Graph plotted with first set of coordinates X and second set Y.
Graph size is set proportional to tanges on X and Y.
Box coordinates:-.5324288   0   -.6306764   0
COORDINATES FILE
C:\Users\Koa\Desktop\EPIC Program Appendix\WORKS Plot\SCO1511.TXT
Graph in file; kepplot.BMP
          COORDINATES

POINT	AXIS 1	AXIS 2
1-	.311-	.631
2-	.283-	.603
3-	.177-	.558
4-	.267-	.389
5-	.249-	.380
6-	.140-	.343
7-	.167-	.185
8-	.138-	.144
9-	.010-	.077
10-	.052	.054
11-	.096	.134
12-	.048	.164
13	.099	.198
14	.165	.254
15	.210	.342
16	.306	.364
17	.251	.449
18	.288	.443
19	.331	.439
20	.287	.469

Intersection of axes is at X=-3.6e-9 AND Y= 9.9999998e-10
Minimum and maximum, horizontal axis:-.5324288 and .5673042
Minimum and maximum, vertical axis:-.6306764 and .4690566
HORIZONTAL AXIS IS AXIS 1
SEQUENCE OF POINTS (LEFT TO RIGHT) IS:
 1 2 4 5 3 7 6 8 11 10 12 9 13 14 15
 17 20 18 16 19
VERTICAL AXIS IS AXIS 2
SEQUENCE OF POINTS (BOTTOM TO TOP) IS:
 1 2 3 4 5 6 7 8 9 10 11 12 13 14 15
 16 19 18 17 20
P4b: kepplot.BMP

3.8.3 Program STEREO

P1: Stereograms are drawn from co-ordinates on three axes stored in an Arrangement 1 file. Data input is by axis. The stereogram pair is stored on in KEP file.

P2: sco1511.txt (same as in Section 3.8.2).
P3: === PROGRAM: STEREO ===
Input data in file:
C:\Users\Koa\Desktop\EPIC Program Appendix\WORKS STEREO\sco1511.txt
Number of data points: 20
Range changed on axes by 0 %
Ranage changed in direction of X,Y,Z by 0 %, 0 %, 0 %
Range of variable 1 = .3210016
Range of variable 2 = .5498665
Range of variable 3 = .21653385
Order on axis 1 :
 1 2 4 5 3 7 8 6 11 10 12 9 13 14 15 17 18 20 16 19
Order on axis 2 :
 1 2 3 5 4 6 7 8 9 10 11 12 13 14 15 16 18 19 17 20
Data
 0 .12675128 .61069803 .1996286 .28023996 .7784368 .65508126 .78774444
 1.3712483 1.1796372 .97988954 1.1991567 1.8660381 2.1698805 2.3745685
 2.8125132 2.5584119 2.7276466 2.92557 2.7247839 0 2.92557 2.92557 0 0
 2.92557 2.92557 0
 0 8.3486264e-2 .21946607 .72524004 .75343188 .86210506 1.3371574
 1.4603238 1.6604797 2.0534573 2.2956483 2.3853381 2.4869942 2.6534115
 2.9176616 2.9846775 3.2399764 3.2210064 3.2099097 3.3 0 0 3.3 3.3 0
 0 3.3 3.3
 0 .69535478 .49890719 .40586424 1.1348889 .83428584 .82602403 1.5002858
 1.4762664 1.0318394 1.4529262 2.0591193 1.4573281 2.4535135 2.3269018
 2.5500373 2.321981 2.2287827 2.3069487 2.80692 0 0 0 0 2.80692
 2.80692 2.80692 2.80692

SPREAD COEFFICIENTS
 2.92557 3.3 2.80692

VIEWPOINTS
 1.287 2.112

1.287 1.287
10.89 9.9

STEREO COORDINATES

0   5.5
4.7613219e-2   5.4913417
.57822676   6.0386161
.15753403   5.6255964
.16311563   5.5671369
.73624317   6.1677959
.60321507   6.0355013
.70797353   6.0761553
1.3844602   6.7550833
1.1683997   6.5820481
.932607   6.3055905
1.1786741   6.4863067
1.9554983   7.3280375
2.4266414   7.6867135
2.6701   7.9459178
3.278956   8.526703
2.9029716   8.179392
3.0983668   8.3860704
3.3659841   8.644241
3.2240669   8.4375785
0   5.5
2.92557   8.42557
2.92557   8.42557
0   5.5
-.44692197   4.7665896
3.4945777   8.7080893
3.4945777   8.7080893
-.44692197   4.7665896
0   0
1.3971858e-3   1.3971858e-3
.1682106   .1682106
.70349306   .70349306
.6913577   .6913577
.82685308   .82685308
1.3412742   1.3412742
1.4880174   1.4880174
1.719049   1.719049
2.1336813   2.1336813
2.4509392   2.4509392
2.6414405   2.6414405
2.6723908   2.6723908
3.0507936   3.0507936
3.360771   3.360771
3.5037614   3.5037614
3.769244   3.769244
3.7186824   3.7186824
3.7267485   3.7267485
3.9990318   3.9990318

```
0          0
0          0
3.3        3.3
3.3        3.3
-.44692197    -.44692197
-.44692197    -.44692197
3.9990318    3.9990318
3.9990318    3.9990318
```

Stereogram file:kepstr.BMP
20090924   09:05:23

P4: The stereo pair below is for stereo viewing at fixed scale. See explanations under P5 in section 3.8.4.

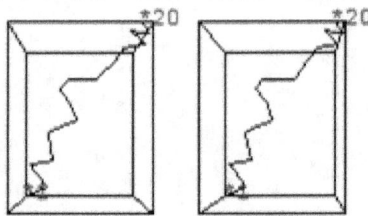

3.8.4 Program TREE

P1: A dendrogram is drawn from data stored on file (TREE) created by program SSA, ALC or SLC. The graph is stored in file KEP. The

P2: TREESSA.txt (created in SSA from raw data EX41221.txt)

P3:   PROGRAM: TREE
======================
DENDROGRAM FILE: treessa.txt
NUMBER OF OBJECTS: 7
HORIZONTAL SEQUENCE:
----
  1 5 2 3 4 6 7
DISSIMILARITY VALUES (VERTICALLY FROM TOP DOWN):

.500   3.333   9.000   10.000   66.400   247.429

P4: kepssa.BMP

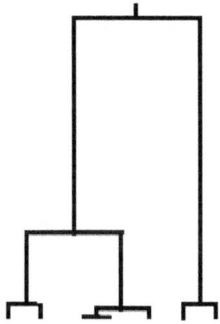

3.9 Multiscale trajectory Analysis

The total analysis is structured by individual tasks, many ways connected, facultative or compulsory. The tasks include: adjustment of the time steps' width, computation of trajectory descriptors (distance, angle, velocity, acceleration, entropy, information, sum of squares), computation of the shape complexity of graphs representing the trajectory or the trajectory properties, computation of the correlation of the trajectory properties and ambient variables, and determination of the concordance (parallelism) of trajectories.

3.9.1 Program BESURIT

P1: Program BESURIT transforms the time series to equal time steps by stretching and summation. Two methods are offered. Method 1 places the mean X=DX/Dt of time interval Dt into the interval Dt times. Then it performs summation of specified number of consecutive units to obtain a new data series of constant time steps. Method 2 keeps X of time point t and defines the next X in the interval as X+(t+1)(DX/Dt). Two time series are input from two files X and t or from a single file. In single file input t is on top. The data are input and output in order from present to past.

P2: The density file testdensities.tru is 3 (taxa) x 4 (time steps):
0
15
21
63
1
30
12
50
2
3
45
6

The ages file testages.tru is 4 valued:
0
5
8
15

P3: Start up dialogue that produced the printda file PRINTDATEST1.TRU below is:
key
Y
testdensities.tru
Y
testages.tru
3
4
3
test1
1
Data1
2
2

P4: selections from the printda file:

Program BESURIT transforms the time series to equal time steps by stretching and sum-mation. Two methods are offerred. Method 1 places the mean X=DX/Dt of time interval DtF into the interval Dt times. Method 2 keeps X of time point t and defines the next X in the interval as X + (t+1)(DX/Dt). Two time series are input in two files
X and t or in a single concottenated file T&X. In single file input t is on top.The data are input andoutput in order from presentto past. The new data set is adjust ted to original variable totals.

Input density file name:
D:\... TRAJECTORY\WORKS BESURIT\testdensities.tru
Input time points file name:
D:\... TRAJECTORY\WORKS BESURIT\ testages.tru

Output file name: data1test1.TRU
Printda file name: PRINTDATEST1.TRU
Actual time points:
 0 5 8 15
Number of variables (e.g.taxa): 3
Number of observations per variable: 4
Input matrix X
 0 15 21 63
 1 30 12 50
 2 3 45 6

Character totals
99 93 56

Streched series
 3 3 3 3 3 2 2 2 6 6 6 6 6 6 6
5.8 5.8 5.8 5.8 5.8 6 6 6 5.4285714 5.4285714 5.4285714 5.4285714 5.4285714
5.4285714 5.4285714
.2 .2 .2 .2 .2 14 14 14 5.5714286 5.5714286 5.5714286 5.5714286 5.5714286
5.5714286 5.5714286
Series total:
 63 85. 82.

Number of cells in streached series: 15
Step size for condensation: 3
Unadjasted data out:
 9 8 10 18 18
 17.4 17.6 17.428571 16.285714 16.285714
 .6 14.4 33.571429 16.714286 16.714286
Adjusted data (e.g. 9*99/63=14.142857):
 14.142857 12.571429 15.714286 28.285714 28.285714
 19.037647 19.256471 19.068908 17.818487 17.818487
 .4097561 9.8341463 22.926829 11.414634 11.414634
Adjuted ages fie:
 0 3 6 9 12
Output vectors: 3 taxa, 5 valued each.

3.9.2 Program STREVOL

P1: This program computes the serial (time, gradient, geo-progression) evolution of structural variables, including compositional distance, compositional transition velocity, acceleration, Rényi's entropy and information, and sum of squared deviations from random expectation, Cordinates and age records are entered. Both, past to present and present to past arrangements are accepted. The output is ordered from past to present! Expectations for mean and variance are found by Monte Carlo simulation under the assumption of chance compositional transitions. Randomisation is within rows or columns depending on the option chosen. Computation of probabilities are based on a normal assumption. When the angular option is selected, all data points are projected onto a unit radius sphere as first step in the computations.

P2: The density file data1test1.tru is 3 (taxa) x 5 (time steps):
14.142857
12.571429
15.714286
28.285714
28.285714
19.037647
19.256471
19.068908
17.818487
17.818487
.4097561
9.8341463
22.926829
11.414634
11.414634

The ages file agefiltest1.tru is 5 valued:
0
3
6
9
12

P3:Start up dialogue that produced the printda file PRINTDATEST1.TRU below is:
N
2
Y
1
R
N
3
5
100
Data1test1.tru
Agefiltest1.tru

P4:Output file VARBSGRAPHSTEST1.TRU contains the graph co-ordinates for
        1 distance
        2 velocity

3 accelleration
4 entropy
5 information
6 sums of squares

Numers 1 to 6 indicate columns in the coordinates file:
12 , 1 , 0 , 0 , 0 , 1.039 , 0 , 45.229
 9 , 2 , 17.091 , 5.69731 , 1.8991063 , 1.039 , 3.615 , 45.229
 6 , 3 , 13.465 , 4.48864 ,-.4028927 , 1.087 , 2.691 , 156.641
 3 , 4 , 9.557 , 3.18566 ,-.43432385 , 1.064 , 4.157 , 30.014
 0 , 5 , 0 , 0 , 0 , .82 , 0 , 91.538
The arrangement is from past to present.

The printdatest1.file contains detailed results and materials for probabilistic inter-
pretations. Some extracts:

Data file: D:\Projects in progress\Eniko\WORKS TRAJECTORY\WORKS
TRAJDESCRIPTION\DATA1TEST1.TRU
Number of variables (dimensions): 3
Number of releves (time points): 5
Ages file:
D:\Projects in progress\Eniko\WORKS TRAJECTORY\WORKS
TRAJ DESCRIPTION\AGEFILTEST
1.TRU
Graphs in file:Varbsgraphstest1.tru
This has 5 rows and 6 columns.
Probabilities are in file:Probsgraphstest1.tru
This has the probability points and probabilities in
alternating columns. Each columns has: 100 elements.
Rows in matrix PARAMETERS
1 mean current
2 cumulant of means
3 cumulant of squared means
4 variance current
5 cumulant of variances
6 cumulant of squared variances
7 observed means
8 observed variance
9 Z mean
10 Z variance
11 Prob mean
12 Prob variance

Time before present
 12 9 6 3 0
Sums of squared deviations from expected mean:
 45.229 , 45.229 , 156.641 , 30.014 , 91.538 ,
Probability graph in file:Probsgraphstest1.tru
Z from .15664168 to 156.64168
Observed mean= 73.730739 Code 7
Iterated mean= 1071.3591 Code 2
Expected mean= 0
Iterated variance of the mean= 11485.396 Code 3

Z=-9.3088456 Code 9
Probability of a more extreme value= 0
Observed variance: 2681.7834 Code 8
Iterated mean variance: 162293.96 Code 5
Iterated variance of variances 1.6684353e+9 Code 6
Bias:-1.6684326e+9 Code 8
Z=-3.9076112 Code 10
Z=-9.3088456 Code 9
Probability of a more extreme value= 0

Distances:
 0 , 17.091 , 13.465 , 9.557 ,
Probability graph in file:Probsgraphstest1.tru
Z from .15664168 to 156.64168

Observed mean= 8.0229775 Code 7
Iterated mean= 12.54925 Code 2
Iterated variance of the mean= 3.3216099 Code 3
Z=-2.4835129 Code 9
Probability of a more extreme value= 0
Observed variance: 60.740406 Code 8
Iterated mean variance: 96.990526 Code 5
Iterated variance of variances 1116.319 Code 6
Bias:-1055.5785
Z=-1.084964 Code 10
Z=-2.4835129 Code 9
Probability of a more extreme value= 0

Velocity values:
 0 , 5.69731 , 4.48864 , 3.18566 ,
Probability graph in file:Probsgraphstest1.tru
Z from .15664168 to 156.64168

Observed mean= 2.6743258 Code 7
Iterated mean= 4.1830835 Code 2
Iterated variance of the mean= .36906777 Code 3
Z=-2.4835129 Code 9
Probability of a more extreme value= 0
Observed variance: 6.748934 Code 8
Iterated mean variance: 10.776725 Code 5
Iterated variance of variances 13.781715 Code 6
Bias:-7.0327815 Code 6
Z=-1.084964 Code 10
Z=-9.3088456 Code 9
Probability of a more extreme value= 0

Accelleration values:
 0 , 1.8991063 ,-.4028927 ,-.43432385 ,
Probability graph in file:Probsgraphstest1.tru
Z from .15664168 to 156.64168

Observed mean= .21237795 Code 7
Iterated mean= .15604054 Code 2

Iterated variance of the mean= 1.5316889e-2 Code 3
Z value= .45520981 Code 9
Probability of a more extreme value= 0
Observed variance: .93301064 Code 8
Iterated mean variance: .99601602 Code 5
Iterated variance of variances .43360113 Code 6
Bias: .49940951 Code 6
Z=-9.5682489e-2 Code 10
Z=-9.3088456 Code 9
Probability of a more extreme value= 0

Entropy values (order one):
 1.039 , 1.039 , 1.087 , 1.064 , .82 ,
Probability graphs for structural variables
 in file:Probsgraphstest1.tru
Z from .15664168 to 156.64168

Observed mean= 1.0102704 Code 7
Iterated mean= .75907073 Code 2
Iterated variance of the mean= 8.8368166e-4 Code 3
Z= 8.4502801 Code 9
Probability of a more extreme value= 0
Observed variance: 1.1679391e-2 Code 8
Iterated mean variance: 3.4860624e-3 Code 5
Iterated variance of variances 1.2784286e-5 Code 6
Bias: 1.1666606e-2 Code 6
Z= 2.2915118 Code 10
Z=-9.3088456 Code 9
Probability of a more extreme value= 0

Information divergenes (order one):
 0 , 3.615 , 2.691 , 4.157 ,
Probability graph in file:Probsgraphstest1.tru
Z from .15664168 to 156.64168

Observed mean= 2.092871 Code 7
Iterated mean= 1.8461741 Code 2
Iterated variance of the mean= .21635653 Code 3
Z= .53036955 Code 9
Probability of a more extreme value= 0
Observed variance: 3.9249553 Code 8
Iterated mean variance: 3.7540457 Code 5
Iterated variance of variances 4.813615 Code 6
Bias:-.88865972 Code 6
Z= 7.7898767e-2 Code 10
Z= .53036955 Code 9
Probability of a more extreme value= 0

3.9.3 Program TRAJCO

P1: Trajectories are compared in pairs by the topological coefficient which is a measure of directional concordance of trajectory pairs. Neigbouring points (Xt and Xt+1,..) are taken in chronological order. Negihbouring points on the axes, are considered identical if a tolerance limit set onto point Xt captures point Xt+1. The tolerance limit is a percent of the coordinate value. Directional concordance is indicated for trajectories a and b when Xat-Xat+1 and Xbt-Xbt+1 has the same sign (++ or --). An equal number of variables is assume in all trajectories. The trajectory lengths may be different. Time points must match between the trajectories. The comparison involves the common portion of the trajectories from present backwards. Coordinates of trajectory points are the input data. Data embedding: trajectory, variable, points. These may be arranged from past to present or present to past. Monte Carlo simulation creates the probabilistic limits and expectations .

P2: The test data is in file testdata.txt whichh includes 6 trajectories describbed by 2 variables and 8,8,8,8,8,6 time points (relevés) each:

1
2
3
4
5
6
7
8
1
2
3
4
5
6
7
8
1
2
3
4
5
6
7

8
1
2
3
4
5
6
7
8
8
7
6
5
4
3
2
1
8
7
6
5
4
3
2
1
4
8
2
6
3
1
5
7
4
5
4
2
7
1
8
6
12
38
31
59
74
91
26
45
28
27
23
26

```
13
25
21
48
42
68
90
120
14
21
7
20
11
10
9
10
```

P3:Start-up dialogue:

```
Consider these before proceding:
================================
-- Coordinates of trajectory points are the input data.
-- The point sequence is the same in all trajectories.
-- The comparison involves the common portion of
   the trajectories from present backwards.
-- The number of coordinate sets is equal in alltrajectories!
-- Data embedding: trajectory, variable, points.
-- Time point matching assumed between the trajectories.
Type 1 to continue, 2 to stop: 1
D:\Projects in progress\ ··· \testdata.TXT
Specify output file name extension:testdat62
Two options are available for arrangement of input data:
     1. first entry in data from top horizon of sediment profile.
     2. first entry in data from bottom horison of sediment profile.

Specify the option -- type 1 or 2: ? 2
Should data ranks be used -- Y/N: y
Maximum lag to shift trajectories:0
Number of trajectories: 6
Number of coordinate sets defining a trajectory: 2
Number of iterations required: 100
Increment in tolerance limit %: 1
Should the curves be put in separate graphs -- Y/N:y
Specify the number of time steps in the trajectories:
Trajectory 1 :? 8
Trajectory 2 :? 8
Trajectory 3 :? 8
Trajectory 4 :? 8
Trajectory 5 :? 8
Trajectory 6 :? 6

Tolerlim % = 20   Iteration: 86
```

P4: Results shown for the 5th and 6th trajectories:
Columns in table: 1- tolerance limit %, 2- expected TC, 3- iterated variance, 4- 95% LL, 5- 95% UL, 6- lag, 7- observed TC, 8- obs-exp, 9- porobsbility of a deviation as large as observed-expected TC.

Traj pair: 5 6

0	.5142	.0249	.2107	.8177	0	.9	.3857	.0166
1	.5106	.0278	.19	.8313	0	.9	.3893	.018
2	.513	.0263	.2011	.8248	0	.9	.3869	.0213
...								
20	.4039	.0239	.1068	.701	0	.8	.396	.0193
...								
100	1	0	1	1	0	1	0	1

(Complete set in PRINTDA file.)

maxima:

20	.4039	.0239	.1068	.701	0	.8	.396	.0193

minima:

0	.5142	.0249	.2107	.8177	0	.9	.3857	.0166

Graph axes: X- tolerance radius %, Y- tpopological coefficient (TC) in first graph and deviation of TC from expectation.

Graphs: a- observed TC, b- upper 95% confidence limit from Monter Carlo simulation, c- expected TC from Monte Carlo simulation,d- lower 95% confidence limit, vertical axis TC in left Graph, and deviation from random expectation in the right hand graph, horisotal axis tolerance radius:

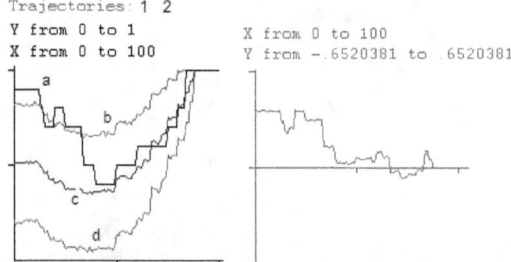

Conclusion: *TC* values are significantly different from random expectation where the graph is outside the *b,c* limits. The deviation is maximal (see table) at tolerance radius 20.

3.9.4 Program LINELENGTH

P1: Program computes graph length and length related fractal dimension in the manner of Mandelbrot's English coast line example. Input data matrix is a series of p x n elements (p variables (taxa), n releves (data or time points)). Adjustment of releve maxima to 100 is an option. Curve length is computed at varying calliper settings r, 2r, 4r, ... When the data contains negative values, data are adjusted to 0 minimum in each taxon. Any number of data sets are allowed, each with its specific p and n.

P2: Input data data2x8.tru:
1
2
3
4
5
6
7
8
12
38
31
59
74
91
26
45

P3: The startup dialogue that produced the output below is:

Key
File name
1
2
8
Y
Y
1
1
2
300
Test1
Data

True BASIC Silver Edition

File

Version 101012 Program computes graph length and length related fractal dimension in the manner of Mandelbrot's English coast line example.
Input data matrix has p x n elements (p variabless (taxa), n releves (data points))
Adjustment of paleoreleve maxima to 100 is an option. Curve length is computed at varying caliper settings r. 2r. 4r. ... When the data contains negative values. data adjusted to 0 minimum in each taxon.

Any number of data sets are allowed. each with its specific p and n. Be prepared

```
to specify ve number of sets, p, and n.
Press a key to continueD:\Projects in progress\Eniko\WORKS TRAJECTORY\WORKS
LINELENGTH\DATA2X8.TRU

Specify the number of data sets:1
Specify the number of taxa and number of releves per data set:
Data set 1 :
Number of variables=? 2
Number of releves=? 8
Number of iterations=100
Do you want to adjust maxima to 100 -- Y/N:y
Do you want intermediate results printed -- Y/N:y
Randomization options: permutation (1) or assortment of totals (2):1
Specify initial caliper width r (say 1):1
Specify constant growth factor in caliper width (say 2):2
Specify upper limit for caliper width:300
Specify output file name extension:test1
Specify output file name:data
  101
Check Printds and Output files for details.
Press a key to exit:
```

P4: Data sets (adjusted to 100 maximum (:
X file
12.5
25
37.5
50
62.5
75
87.5
 100
Y file
13.18681319
41.75824176
34.06593407
64.83516484
81.31868132
100
28.57142857
49.45054945

For these the Y over X graph is

124

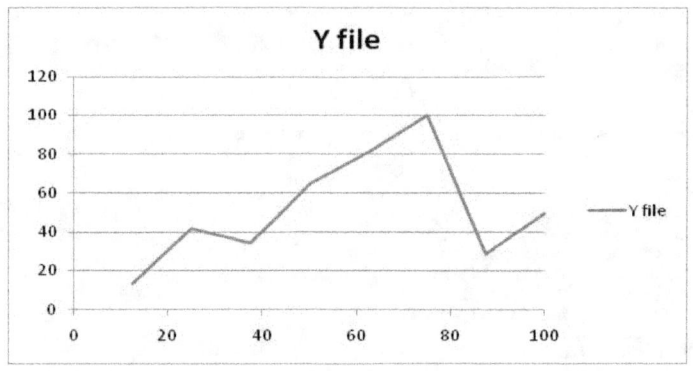

The printdatest1.tru file has much detail about the run. File datatest1.tru has a summary of the evolution of D as r is incremented in steps. The column headings:
r, L(r), n (number of r values in regression), b, D, Mean D
Data set: 1
****
Graph #: 1
Original length= 40.043419

r	L(r)	n	b	D	Mean D
1	19				
2	14	2	-0.44057	1.440573	1.440573
3	18	3	-9.14E-02	1.091356	1.265965
4	12	4	-0.23744	1.237435	1.256455
5	10	5	-0.30005	1.335748	1.276278
6	12	6	-0.30005	1.300049	1.281032
7	14	7	-0.23031	1.230313	1.272579
8	16	8	-0.15532	1.155316	1.255827
9	18	9	-8.40E-02	1.083987	1.234347
10	10	10	-0.1324	1.132396	1.223019
11	11	11	-0.14979	1.149789	1.215696
12	12	12	-0.1491	1.149103	1.209642
13	13	13	-0.13774	1.13774	1.203651
14	14	14	-0.12009	1.120093	1.197223
15	15	15	-9.88E-02	1.098844	1.190196
16	16	16	-7.57E-02	1.075664	1.182561
17	17	17	-5.16E-02	1.051608	1.174376
18	18	18	-2.73E-02	1.027345	1.165727

The reliability of D increases with increased n.

The regression results in last line (r=18) for b and D are bades on

ln r	ln L
0	2.944438979
0.693147181	2.63905733
1.098612289	2.890371758
1.386294361	2.48490665
1.609437912	2.302585093
1.791759469	2.48490665
1.945910149	2.63905733
2.079441542	2.772588722
2.197224577	2.890371758
2.302585093	2.302585093
2.397895273	2.397895273
2.48490665	2.48490665
2.564949357	2.564949357
2.63905733	2.63905733
2.708050201	2.708050201
2.772588722	2.772588722
2.833213344	2.833213344
2.890371758	2.890371758

The regression equation is $L(r) = -0.0273 \ln(r) + 2.7021$ and $D = 1-(--0.0273) = 1.0273$. This D is close to the minimum, implying a very simple graph shape.

Based on the probability values found at the end of the printdatest1.tru file, our conclusion is: the very high negative z value reinforces the above conclusion that the complexity level is lower than could be expected with random transitions, the deviation from zero is still highly significant.

3.9.5 Program SLIDECORR

P1: The sliding windows correlations program (Method 1) computes regression based estimates for correlations and correlation signe frequencies for one trajectory descriptor (e.g. velocity) and a forcing variable (e.g. temperature). Input includes single column data files of two types:

FILE 1. Finger TBP series concattenated with the observed trajectory descriptor. The 'finger' series assumes that the TBP series in the forcing factor source is more detailed than the observed TBP series.

FILE 2. Reference TBP is concatenated with the forcing variable from any source. Frequencies of positive and negative correlations are generated by sampling and resampling of the series a large number of times using randomly sited windows of random length (min 5 time steps).

P2:data set TBPVELOC132.TRU – 132 time records followed
by 132 velocity records:
587
1148
1418
1683
2194
...
34797
36288
37841
39457
41138
0.33877
0.89563
2.07628
1.79137
1.08263
...
0.04325
0.06443
0.02714
0.02428
0.03528

VOSTOK3311.TRU 3311 time records followed by 3311
temperature records:
0
17
35
53
72
91
110
...

420281
420888
421507
422135
422766
0
0
0
0
0
0
0
0
-0.81
0.02
0.36
...
0.51
0.54
0.32
0.15
0.08
0.23
(See complete data sets with application program.)

P3: Startup conversation:

```
True BASIC Silver Edition
The sliding windows correlations program (Method 1) computes regression
based estimates for correlations and correlation signe frequencies for
one trajectory descriptor (e.g. velocity) and a forcing variable
(e.g. temperature). Input includes single column data files of two types:
FILE 1 fingers TBP series concatenated with the observed trajectory
descriptor. The 'finger' series assumes that the TBP series in the
forcing factor source is more detailed than the observed TBP series.
FILE 2 contains reference TBP concattenated with forcing variable from any
source.
Frequencies of positive and negative correlations are generated by
sampling and resampling of the orginal series a large number of times
using randomly sited windows of random length (min 5 time steps).

Press a key to continue:
D:\Projects in progress\Eniko\WORKS TRAJECTORY\WORKS SLIDECORR\TBPVELOC132.TRU
D:\Projects in progress\Eniko\WORKS TRAJECTORY\WORKS SLIDECORR\VOSTOK3311.TRU
Specify the full length of the structure variable series:132
Specify the full length of the forcing vector series:3311
Specify the number of iterations:100
Specify maximum lag:1

1  40  100
Run completed. Press a key to exit.
```

P4: Trajectory descriptors related to the correlation of Vostok temperatures and Hanging Lake compositional transition velocity readings. Table heading:
Lag, BS (window size), r(V,T), F- (frequency of negative correlations in sampling and resampling), F+ (frequency of positive correlations:

```
0  1  .49   22  74
0  2  .557  24  74
0  3  .616  25  74
0  4  .646  26  72
0  5  .663  25  74
0  6  .678  27  72
0  7  .696  23  74
0  8  .713  20  78
0  9  .727  31  69
0  10 .739  25  75
0  11 .753  35  64
0  12 .766  24  74
0  13 .776  22  77
0  14 .784  26  73
0  15 .792  26  74
0  16 .801  24  75
0  17 .809  36  63
0  18 .817  31  68
0  19 .824  21  78
0  20 .832  30  70
0  21 .84   22  77
0  22 .849  20  79
0  23 .857  24  73
0  24 .865  26  74
0  25 .873  22  77
0  26 .882  10  90
0  27 .89   30  68
0  28 .897  17  79
0  29 .905  16  84
0  30 .91   17  82
0  31 .915  21  77
0  32 .919  26  73
0  33 .922  28  69
0  34 .926  18  78
0  35 .931  21  76
0  36 .936  22  76
0  37 .939  24  72
0  38 .943  15  81
0  39 .947  18  79
0  40 .95   16  81
```

Eq1 .623 .0093
Eq2 27.407 -.1955
Eq3 71.694 .1576

Regression equations at increased lag (temperature records shifted up in steps of 1)from 0 to 40:

$r(V,T)$ F-   F+

Lag	a1	b1	Y	a2	b2	Y	a3	b3	Y
0	0.62	0.01	0.63	27.41	-0.20	27.21	71.69	0.16	71.85
1	0.62	0.01	0.63	32.99	-0.48	32.51	66.68	0.40	67.09

Conclusion:
a. r(V,T) and F+ decline slightly with increasing lag. F- rises with increasing lag.
b. F+ dominates. This is characteristic for the humid and wet zones. The opposite is expected in arid zones.

3.10 Markov chain analysis

P1: Serial data are expected and constant time step width is assumed. Transition probabilities are computed from relevé level population gains and losses across steps with LAG 1,2, ... and up . The stress of the distance structure of natural relevés and Markov relevés are compared based on the Kruskal stress index. Probabilities are derived by randomization in Monte Carlo experiments. The randomization is implemented either by permutation of relevé positions in the series (option 1) or random resampling of the stretched Markov chain (option 2).

P2: The data file testdat.tru is 3 (taxa) x 4 (equidistant stesps):
46.7222
33.4027
19.8751
27.3256
36.3372
36.3372
16.6667
25
58.3333
0
50
50

P3: Start-up dialogue:

```
The MARKOV chain is fitted to an observed series. Options for
randomization testing of the closeness of fit are included. Input
data arranged by taxa or releves in a single column. Constant
time-step width is assumed.
Do you wish to proceed? -- press Y or N:y

D:\Projects in progress\ \Markov\TESTDAT.TRU
Should intermediate results be stored -- press Y or N:y
Data arranged by taxa?-press Y or N:n
Specify number of taxa: 3
Specify number of releves (length of the series): 4
Specify LAG size upper limit 0,1,2, .:2
```

```
Choose hypothesis to be tested:
 1- Ho: series is 0-order Markov (undirected); random permutation of positions
 2- Ho: series is m-order Markov; sampling/resampling of the stretched M
Choose 1 or 2: 2
Specify % error threshold:1
Output file name extension: outlag2rnd2
STEP SIZE: 1
=========================================
Randomization begins. Specify number of iterations to be used:? 100


Iteration #: 100
STEP SIZE: 2
=========================================

Iteration #: 100
Printda file:printdaoutlag2rnd2.tru
Consult instruction at end of Printda file.
Press any key to exit.
```

P4: Edited PRINTDA file:

PROGRAM: FitmarkoR.tru
Key reference:
Orlóci, L., Anan, M. and X.S. He. 1993. Markov chain: a realistic model for temporal coe-nosere? Biometrie-Praximetrie 33:7-26.
===
Data file: D:\Projects in progress\UFRGS Statistical Ecology & Short course\UFRGS Traj & Au to corr\Markov \TESTDAT.TRU This has 3 taxa and 4 releves. Hypothesis tested: Ho: series is 2nd -- order Markov; sampling/resampling of the stretched M used.

Output files: Stepwise transitions in file Trnsstepoutlag2rnd2.tru; this has 4 sets of 3 by 3 numbers.

Global transitions in file Trnsproboutlag2rnd2.tru; this has 3 rows and 3 columns.

Markov relevés in file Markdatoutlag2rnd2.tru; this has at least 4 rows and 3 columns.

Transposed Markov relevés in file Tmarkdatoutlag2rnd2.tru; this has 3 rows and 4 columns.

Step size upper limit used: 2

Data:
```
 46.7222   27.3256   16.6667    0
 33.4027   36.3372   25        50
 19.8751   36.3372   58.3333   50
```
Column totals:
```
 100      100      100      100
```

STEP SIZE: 1
===
Stepwise transitions (rows, present; columns, future):
Step 1
```
  .55202  .14395  .30403
  .00000  .81967  .18033
  .00000  .02913  .97087
```

Step 2
```
.60837  .07985  .31179
.04674  .68160  .27167
.00000  .00000 1.00000
```
Step 3
```
.00000  .51376  .48624
.00000 1.00000  .00000
.00000  .28716  .71284
```
Averaged transitions (rows, present; columns, future):
```
.46235  .19639  .34126
.01167  .86703  .12130
.00000  .12817  .87183
```
Markov analysis
Startup releve: 1
```
 46       27       21
```


Generated releves
Releve: 2
```
 21.583211  35.135254  37.281535
```
Instability level at this step= 30.454069 or 100 %
Instability reduction= 0 %
Releve: 3
```
 10.389001  39.480366  44.130633
```
Instability level at this step= 13.82391 or 45.392655 %
Instability reduction= 54.607345 %
Releve: 4
```
 5.2640424  41.927131  46.808827
```
Instability level at this step= 6.2788991 or 20.617603 %
Instability reduction= 24.775052 %

...
Releve: 10
```
 1.0247618  45.823171  47.152068
```
Instability level at this step= .27135346 or .89102532 %
Instability reduction= .30471289 %

Distance structure and randomization experiment
Two sets of distances are computed, one for the Markov relevés and the other for the natural relevés. The comparison is based on the Kruskal's stress coefficient. Significance of the stress value is determined in randomization testing.
Number of iterations requested: 100
Transposed Markov releves in file Tmarkdatoutlag2rnd2.tru
Distance file created: Distoutlag2rnd2.tru
Number of elements per distance set: 6
Output file name extension: outlag2rnd2
Number of iterations: 100
Distance matrix for X:
```
 29.883106  48.010416  63.095166
 20.149442  38.105118
 38.807216
```
Distance matrix for M
```
 30.454069  44.25979  50.481016
 13.82391  20.080193
 6.2788991
```

Stress of observed and Markov structures = 0.44935669
Stress values straight and and in standard units =
(1) 7.0918256e-2 ,-21.961793 (2) 8.0587915e-2 ,-20.501723 (3)
8.4100092e-2 ,-19.971402 (4) 9.3682077e-2 ,-18.524571 (5) 9.5276013e-2 ,
-18.283894 (6) 9.8559062e-2 ,-17.788171 (7) .10115145 ,-17.396733 (8)
.11349922 ,-15.532282 (9) .11505879 ,-15.296796 (10) .11966874 ,
-14.600716

...

(90) .30259123 , 13.019655 (91) .30534756 , 13.435848 (92) .31353216 ,
14.671682 (93) .31894058 , 15.488325 (94) .3216473 , 15.897026 (95) .33275419 ,
17.57411 (96) .3354182 , 17.976362 (97) .33614864 , 18.086655 (98) .34464511 ,
19.369579 (99) .34829485 , 19.920672 (100) .35029028 ,
20.221971
Observed stress value: 0.44935669
Percent of generated values at least as large as 0.44935669 is 0
Mean of generated stress values: 0.21636546
Variance of the generated stress values: 4.3860661e-3
Standard deviation of the generated stress values: 6.6227382e-2
Variance of the generated mean: 4.3860661e-5
Standard deviation of the generated mean: 6.6227382e-3
Observed stress expressed as a deviation from the generated mean
in standard units: 3.5180499

Test of Ho: the coenosere is m-order Markov.
Test criterion: s(DX;DM). In this, DX is the distance configuration
of the observed relevés and DM is the distance configuration of
relevés based on the Markov scores M fitted to X. The reference
distribution is based on RNDM, the random-permuted M, and s(DRNDM;DM) but in this
case DM is the Markov distance configuration based on M fitted to RNDM. Probabilities
and probability points are given below. The probabilities, given as % values, are in the
right tail of the
s(DRNDX;DM) distribution.
99%   95% 90%   75%   50%   25%   10%   5$ 1%
.0709 .0953 .11967 .17107 .21355 .26400 .30259 .33275 .34829
*The probability points are stress values. Reject the null hypothesis
when the observed stress value associates with a small probability.

STEP SIZE: 2
==
Stepwise transitions (rows, present; columns, future):
Step 1
 .42331 .06643 .51026
 .10682 .38804 .50514
 .00000 .00000 1.00000

Step 2
 .00000 .50000 .50000
 .00000 .95238 .04762
 .00000 .09910 .90090

Averaged transitions (rows, present; columns, future):
 .30916 .18335 .50749
 .05324 .67114 .27562
 .00000 .03928 .96072

Markov analysis
Startup releve: 1
 46   27   21

Startup releve: 2
 21.583211 35.135254 37.281535

Generated releves
Releve: 3
 15.658544 27.37991 50.961546
Instability level at this step= 16.804456 or 56.068956 %
Instability reduction= 43.931044 %

Releve: 4
 8.5430439 29.00244 56.454516
Instability level at this step= 9.1343132 or 30.477119 %
Instability reduction= 25.591837 %

Releve: 5
 6.2985334 23.248542 64.452924
Instability level at this step= 10.105429 or 33.717299 %
Instability reduction=-3.2401797 %
 ...

Releve: 13
 1.2227898 12.946753 79.830457
Instability level at this step= 1.7686579 or 5.9012208 %
Instability reduction=-4.6567416 %

Releve: 14
 1.185908 12.800585 80.013507
Instability level at this step= .23713453 or .79121191 %
Instability reduction= 5.1100088 %

Distance structure and randomization experiment

Two sets of distances are computed, one for the Markov releves
and the other for the natural releves. The comparison is based on
the Kruskal stress coefficient. Significance of the
stress value is determined in randomization testing.

Transposed Markov releves in file Tmarkdatoutlag2rnd2.tru
Distance file created: Distoutlag2rnd2.tru
Number of elements per distance set: 6
Output file name extension: outlag2rnd2
Number of iterations: 100
Distance matrix for X:
 31.400637 62.64982 56.789083
 36.783148 29.580399
 46.281746
Distance matrix for M
 30.454069 42.6432 51.614495

16.804456 23.984591
9.1343132
Stress of observed and Markov structures = .50237306
Stress of random and Markov structures =
(1) 7.1687613e-2 ,-17.245463 (2) 7.2370396e-2 ,-17.170398 (3) 7.8798693e-2 ,-16.463663 (4) 7.9451542e-2 ,-16.391888 (5) 8.6523919e-2 , -15.614343 (6) .10603245 ,-13.469552 (7) .10754648 ,-13.303098 (8) .11283647 ,-12.72151 (9) .11363862 ,-12.63332 (10) .11622989 ,-12.348433

....

(90) .34307447 , 12.591127 (91) .35352616 , 13.740198 (92) .36727184 , 15.251414 (93) .37833955 ,16.468211 (94) .37976677 , 16.625122 (95) .38206807 , 16.87813 (96)
.4122748 , 20.199093 (97) .41812738 , 20.842532 (98) .42371632 , 21.456986 (99) .43416125 , 22.605315 (100) .5025034 , 30.118931
Observed stress value: .50237306
Percent of generated values at least as large as .50237306 is 1
Mean of generated stress values: .22854844
Variance of the generated stress values: 8.2733089e-3
Standard deviation of the generated stress values: 9.0957731e-2
Variance of the generated mean: 8.2733089e-5
Standard deviation of the generated mean: 9.0957731e-3
Observed stress expressed as a deviation from the generated mean
in standard units: 3.0104602

Test of Ho: the coenosere is m-order Markov.
Test criterion: s(DX;DM). In this, DX is the distance configuration
of the observed relevés and DM is the distance configuration of
relevés based on the Markov scores M fitted to X. The reference
distribution is based on RNDM, the random-permuted M, and s(DRNDM;DM)
but in this case DM is the Markov distance configuration based on M
fitted to RNDM. Probabilities and probability points are given below.
The probabilities, given as % values, are in the right tail of the
s(DRNDX;DM)distribution.
99%   95%   90%   75%   50%   25%   10%   5$ 1%
.0717 .08658 .11623 .16063 .2281 .28899 .34307 .38207 .43416
*The probability points are stress values. Reject the null hypothesis
when the observed stress value associates with a small probability.

Test of Ho: the coenosere is m-order Markov.
Test criterion: s(DX;DM). In this, DX is the distance configuration
of the observed relevés and DM is the distance configuration of
relevés based on the Markov scores M fitted to X. The reference
distribution is based on RNDM, the random-permuted M, and s(DRNDM;DM)
but in this case DM is the Markov distance configuration based on M
fitted to RNDM. Probabilities and probability points are given below.
The probabilities, given as % values, are in the right tail of the
s(DRNDX;DM)distribution.
99%   95%   90%   75%   50%   25%   10%   5$ 1%
.0579 .1025 .11522 .13668 .20998 .25990 .30437 .32902 .43308
*The probability points are stress values. Reject the null hypothesis
when the observed stress value associates with a small probability.

Glossary

Executable program: code translated into machine language and linked with all needed library functions to stand alone as an .EXE application.

Dynamic array: its dimensions expand as needed in the program.

Individual: the unit object on which variables are measured; a relevé which describes the object; a vector (relevé) whose elements are kept conterminous in the data file.

Order of a symmetric matrix: the number of rows (columns) in the matrix.

Primary data file: a user-created data file.

Random access: data positions accessible in any sequence; adjacency does not rule the sequence of progress in reading from it or writing onto it.

Rank of a matrix: latent dimensionality, related to the number of non-zero latent roots (Eigenvalues).

Relevé: a record set in which an individual object's description is stored; a description of any kind; a vector whose elements are kept conterminous in the data file.

Sequential file: units are read or written in order starting with the first; no position can be stepped over.

Start-up dialogue: the initial conversation of user and program as the run gets under way.

Variable: a character that varies

Key references

Feoli, E. and L. Orlóci (eds.) 1991. Computer Assisted Vegetation Analysis. Kluwer Academic Publishers, London. -- a broad account of advances in a changing field and exposition of views that characterise the changes.

Morrison, D.F. 1976. Multivariate Statistical Analysis. 2nd ed. McGraw-Hill, New York. – intermediate to advanced level text for general users with familiarity in linear algebra and univariate statistics.

Orlóci, L. 1971. Information theory techniques for classifying plant communities. In: G. P. Patil, E. C. Pielou and W. E. Waters (eds.), Statistical Ecology, Vol. 3, pp. 259-270. Pennsylvania State University Press, University Park.

Orlóci, L. 1978. Multivariate Analysis in Vegetation Research. 2nd ed. W. Junk, The Hague. – a monograph of the field as it was in the 1970s.

Orlóci, L. 1991. Entropy and Information. Ecological Computations Series: Vol. 3. SPB Academic Publishing, The Hague. -- detailed account of techniques, examples of application and a computer program.

Orlóci, L. 1991. CONAPACK: A program for Canonical Analysis of Classification Tables. Ecological Computations Series: Vol. 4. SPB Academic Publishing, The Hague. -- detailed account of techniques, examples of application and a computer program.

Orlóci, L. 1991. On character-based community analysis: choice, arrangement, comparison. In: Feoli, E. and L. Orlóci (eds.) Computer Assisted Vegetation Analysis, pp. 81-93. Kluwer Academic Publishers, London.

Orlóci, L. 1993. The complexities and scenarios of ecosystem analysis. In: G.P. Patil and C.R. Rao, Environmental Statistics, pp.421-430, North Holland/Elsevier, New York.

Orlóci, L. 2010. Statistical Ecology. A reasoned approach. Scada Publishing. Online edition.

Orlóci, L., C. R. Rao and W. M. Stiteler (eds.) 1980. Multivariate Methods in Ecological Work. ICP, Fairland, Maryland.

Orlóci. L and M. Orlóci. 1991. Edge detection in vegetation: Jornada revisited. In: Feoli, E. and L. Orlóci (eds.) Computer Assisted Vegetation Analysis, pp. 373-385. Kluwer Academic Publishers, London.

Pillar, De Patta V. and L. Orlóci. 1993. Character-based Vegetation Analysis: the Theory and an Application Program. Ecological Computations Series (ECS): Vol. 5. SPB Academic Publishing bv, The Hague, The Netherlands. -- arguments in favour of character-based analysis and description of a very fast computer program in C.

Three outstanding ecological program packages by V. De Patta Pillar, J. Podani, and O. Wildi. – find these on the Internet.